Geospace 2060

Geospace 2060

How Earthlings Colonized the Solar System

Walter Gomez

Copyright © 2017 by Walter Gomez
All rights reserved. No part of this publication may be reproduced, distributed, or transmitted in any form or by any means, including photocopying, recording, or other electronic or mechanical methods, without the prior written permission of the author, except in the case of brief quotations embodied in critical reviews and certain other noncommercial uses permitted by copyright law.

ISBN: 1494752026
ISBN 13: 9781494752026

Walter Gomez also has written
THE MILITARY GEOGRAPHY OF THE SOLAR SYSTEM

Dedicated to my life editor, Telecha
&
My brother, Patrick, who edited this book

Acknowledgement

The narratives have been inspired by the work of the many Enlightened Minds that have worked throughout history since the Great Human Awakening Event to advance the Cosmic Imperative of human colonization of the Solar System.

List of Narratives

Acknowledgement · ix
Author's Prologue· xiii

Chapter 1	The Great Human Awakening Event · · · · · · · · · · · · · · ·	1
Chapter 2	The Earth at the Time of The Ghae· · · · · · · · · · · · · · ·	7
Chapter 3	The Tropic of Cancer Region · · · · · · · · · · · · · · · · · ·	11
Chapter 3	Making of the Planetary Colonizer · · · · · · · · · · · · · · ·	13
Chapter 4	Discovering Planet Earth ·	19
Chapter 5	Separate and Superior· ·	23
Chapter 6	Ways and Means of Colonization · · · · · · · · · · · · · · · ·	25
Chapter 7	The First Planetary Colonization · · · · · · · · · · · · · · · ·	31
Chapter 8	The Pleistocene Human Region· · · · · · · · · · · · · · · · ·	41
Chapter 9	The Holocene Colonization · · · · · · · · · · · · · · · · · ·	44
Chapter 10	Human Spatial Systems ·	49
Chapter 11	The Making of the Holocene Planet · · · · · · · · · · · · · ·	53
Chapter 12	The Holocene Human Incubator· · · · · · · · · · · · · · · ·	58
Chapter 13	New Beginnings· ·	62
Chapter 14	The Calculus of Space Colonization· · · · · · · · · · · · · · ·	73
Chapter 15	Proactive Colonization ·	77
Chapter 16	Earthlings in Space ·	82

Chapter 17	The Second Human Awakening Event	88
Chapter 18	The Space Imperative	91
Chapter 19	Geometries of Space	104
Chapter 20	Knowledge-Technology-Skills	115
Chapter 21	Planetary Colonizations	118
Chapter 22	Navigating The Gravity Ocean	121
Chapter 23	Energy is Matter is Mass	130
Chapter 24	"Weather"	134
Chapter 25	Planetary Colonies	157
Chapter 26	Diverse Populations	162
Chapter 27	A Low-Mass Strategy	176
Chapter 28	Orbiting Space Colony Strategy	184
Chapter 29	Terraforming Space Colonies	193
Chapter 30	Creating A Geospace Region	205
Chapter 31	Governing the Geospace Region	209
Chapter 32	The Making of Space Societies	217
Chapter 33	Human Resources on Space Colonies	225
Chapter 34	Geospace Economics	237
Chapter 35	Harnessing the Forces of the Cosmos	242
Chapter 36	The Geospace Network	253
Chapter 37	The Geomars Transportation System	259
Chapter 38	Nanotechnology in Space	263
Chapter 39	Quo Vadis?	268
	Epilog	275

Author's Prologue

It is the year 2162 CE of the human calendar on earth. I am Josef Herzog, a Carbon-Based human; a historian who resides on space colony IPM-23, a node of the network of colonies that lies within the L1 Lagrangian Point of the sun.

The following is a series of narratives that deal with the history of the colonization of the solar system by earthlings, from the time of the Great Human Awakening Event (GHAE) which occurred approximately 2 million years prior to the present.

Each narrative has been written by a chronicler who has lived at various times and at various places on earth. Each narrative presents a topic that is considered most relevant to the overall description of generational events that have contributed to the human advances in actualizing the Cosmic Imperative of colonizing the Solar System.

The record of human thought and activity throughout human history has been compiled in the Cosmic Archive, which opened at the time of the GHAE. It contains the digitized, cumulative memory of all the activities of the human brain and the human mind that have occurred throughout the history of Pleistocene and Holocene colonization of the earth, and of Holocene colonization of the rest of the Solar System. Indeed, the relative swiftness of the exploration and colonization of the entire Solar System has been possible only because of the cumulative knowledge and wisdom that is now available to modern earthlings in the 21st century CE.

The information contained in the Cosmic Archive has been translated into the generally-accepted computer machine language and has been digitized to make it accessible to anyone in the universe. Machine language is the *lingua franca* of all programmable computers; it is the *Rosetta Stone* and the *Enigma Machine* by which raw data and refined information can be converted to this universal language of both carbon-based silicon-based entities.

The Cosmic Archive also contains an integrated digitized search paradigm which enables modern earthlings to quickly access specific information which has already been analyzed and synthesized to respond to any search arguments – virtually in real time. Thus, for instance, it can provide a succinct information on of the history of the human colonization of the Solar System: such as, an analysis of the Pleistocene *Incubation Period*, which took place in the Great Rift Valleys of Africa and which lasted for about 200,000 years, during which time the Pleistocene humans developed the body and brain system that prepared them for the first wave of planetary colonization on Earth. Also, that the first phase of human colonization of the planet Earth ended during the Pleistocene *time of troubles* – the millennia during which the Earth's surface underwent a complete renovation – a realignment of the waters and drylands – following a period of global warming, deglaciation, and the subsequent reallocation of the hydrosphere – from the ice domains to the oceans and into the atmosphere. And, that a second phase of colonization of the home planet began about 30,000 years BCE, and that it coincided with the beginning of the Holocene geological period and the emergence of an advanced group of Holocene humans.

> *[The central theme of this collection of narratives is the unbroken line of contributions by a variety of Enlightened Humans to the ultimate colonization of not only the Earth, but also the rest of the Solar System – as well as those of generations of human groups, including the Pleistocene humans to the Homo sapiens and the silicon-based Homo spaciens… which are also known as humanoid robots or androids.]*

THE STORY BEGINS WITH THE GIFT OF HUMAN AWAKENING… GIVEN TO EARLY PLEISTOCENE HUMANS TWO MILLION YEARS AGO BY THE COSMIC MIND…

Chapter 1

The Great Human Awakening Event

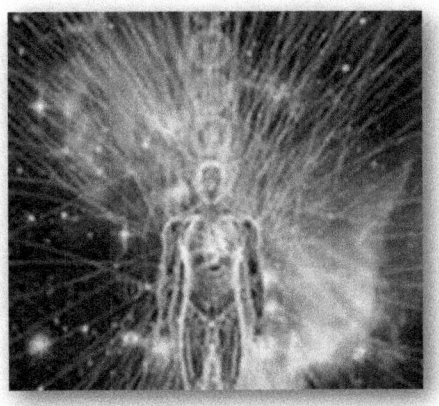

*THE COSMIC MIND ESTABLISHED THE LAWS
THAT GIVE ORDER TO THE COSMIC FIELD...*

Two million years ago, the Cosmic Mind transmitted to a certain group of Pleistocene hominids the spark of humanity which would give them the attributes of self-awareness, relational awareness, and many other powers of the brain that are now ascribed to modern Homo sapiens.

The Cosmic Mind also gave the nascent humans the command to go forth and colonize the Solar System – one planet at a time – beginning with the Earth. The transmission also included a complex of algorithms and protocols that would guide the evolution of humans to prepare them for accomplishing the mission of colonizing the Solar System by the Earth-year 2060 CE.

Towards this end, the Cosmic Mind would give programming instructions that would continue to be transmitted via a series of epigenetic infusions into the DNA of the humans throughout their period of evolution, and continue into the era of the colonization of the Solar System. They first directed the evolution of the brain, body, mind, and culture throughout human history on Earth, and now they guide the evolution of modern Homo sapiens toward *Homo spaciens* – the earthlings that will be able to live and work throughout the Solar System, without the need for massive assistance from technology.

So, something happened approximately 2 million years ago; it set into motion a series of events which would lead to the colonization of the entire Solar System by humans in the middle of the 21st century. Homo erectus was the first to receive the directive programming, and it inspired them to change their strategy for survival, as they descended from the trees and began to dwell on the ground. It would be the first "small step by man and a giant step for mankind…"

So, it was the Great Human Awakening Event (GHAE) which precipitated the decision to not only descend from the trees, but to also take up a bipedal mode of locomotion; to adopt the lifestyle of the roaming hunter and gatherer; to fashion the assortment of tools and to learn the skills for the economic system they had chosen; to organize into efficient social groups for surviving and propagating; and to migrate from one place to another, in search of a new resource site, where water, food, and materials could be found to press on with the colonization of the planet. It was the trajectory along which Pleistocenes would progressively develop larger and more complex brains, and more agile and powerful bodies.

Thus equipped, the Pleistocene humans, almost immediately, began to take the first steps (literally) toward the development of a highly effective *migration and colonization* civilization on the planet. It was this mind-set and behavioral paradigm also would empower Holocene humans in their quest to rove and colonize the rest of the Solar System almost two million years later.

The programmed instructions to humans for colonization are externally deterministic and internally opportunistic. They provide a dynamic and random framework for decision-making, in which the human neocortex is given the "free will" to respond with genetic adaptations to environmental changes. The *Operating System* and the original suite of programs that has guided the development of humans on Earth and in space, occurs at the level of the cell, in which the DNA-RNA system transmits directions and fashions the optimum human system for accomplishing the missions of colonization.

Other, less deterministic, programming instructions are written by humans themselves; they are genetic and epigenetic instructions to the offspring that have driven seemingly random and fortuitous changes in the human organism and culture – but are really mutation-selection events – which have enabled the species to confront and overcome the many cataclysmic events that might otherwise have resulted in the extirpation or even, extinction of the species during the past thousands of millennia on Earth.

On a physical level, programming instructions are blueprints and algorithms for the construction of the humans as optimum planetary rovers on Earth – very much like the artificial planetary rovers that would be sent to the Moon, Mars and the other planets and moons of the Solar System two million years later. In terms of "artificial intelligence," the instructions would serve to develop the powers of the human brain. And, later, they would create the higher consciousness and intuitive reasoning of the human mind.

Walter Gomez

The Anatomy And Physiology Of Programming

Genetic Instructions
    Chromosomes (DNA = carries genetic information)
        Cells (nucleic acids + protein)
            Cells (genes = genetic information)

Genetic instructions are encoded on a set of binary settings within the chromosomes; threadlike structures of nucleic acids and protein that are found in the nucleus of most living cells. Within the chromosomes there is a self-replicating material called DNA (deoxyribonucleic acid) which is the actual carrier of genetic information. The genetic information serves as a blueprint for the architecture and behavior of each cell, which determines the distinctive system of the anatomy and physiology, as well as the "personality" of each new human. More broadly, there is a genetic transfer of DNA that the parents pass on to the child at the time of conception. It is the mechanism by which many of the *hardwired* traits – as well as the cumulative knowledge of the Collective Unconscious – have been passed on to later generations of humans since the GHAE.

Then, throughout the life of the organism, there is a continuing input of external (environment) and internal information from within the organism. These are called epigenetics, to distinguish them from the original genetic information that is passed onto the human by the Cosmic Mind and its parents at the time of conception. The full power of epigenetics is both unique and characteristic in humans. It is what would energize and direct the development of their brain, body, mind, and cultures as they prepared for the ultimate objective of colonizing the Solar System.

And so, to prepare the Pleistocene humans for fulfilling the GHAE mandate of colonizing the Solar System, they undergo many changes in their own organism to deal with both subtle and acute changes in their external environments… over space and time. Therefore, they were given the power to change their own organism; to purposefully respond to external environmental changes – through the duality of mutation and selection… and by cultural adaptation – to a far greater degree than any other carbon-based organism on Earth.

Thus, the Pleistocene colonizers began to adapt organically to changes in the external environment, within only a few thousand generations. Highlanders adapted genetically to prevent a condition in which the body produces too many red blood cells in response to oxygen deprivation, by enhancing the capability of their glucose cells to retain and metabolize the additional oxygen cells. Other early humans developed mutations in their DNA-RNA processes, which enabled them to survive in conditions of crippling heat and humidity, by developing extra sweat glands, for instance. In colder regimes, they grew thicker hair all over their body to retain their core heat. And, they adapted their gastro-intestinal systems to make them virtually omnivorous; sometimes with the help of processing and cooking technologies and techniques. This is how humans have been so successful in adapting to the various regimes which they have encountered.

Chapter 2

The Earth at the Time of The Ghae

At the time of the Great Human Awakening Event, the planet Earth had completed its early period of construction, which began about 4 billion years earlier. By then, the continents, the atmosphere and the hydrosphere had entered relative stasis.

The great internal upheavals that caused the global movements of the tectonic plates had subsided to the point where the continents had settled into their modern positions. This was the geological period known as the Pleistocene epoch; it began at just about the time of the GHAE and coincided with the emergence of the Pleistocenes; the first humans.

However, there was still major remodeling of the African continent; it was still dividing into two parts (east and west), along a longitudinal Great Rift Valley. And, within this system of valleys, escarpments and lakes, a continuing series of volcanic and seismic events would create finishing touches. Meanwhile, in the Great Rift Valleys, the lakes and the grasslands and intermittent forests attracted large populations of animals, thus producing a relative paradise in which the earliest Pleistocenes could develop and propagate. In other terms: it was an excellent incubator for the developing humans on Earth.

The Pleistocene Epoch encompassed both the last Ice Age and the tenure of the Pleistocene humans on Earth. On a global scale, the Pleistocene epoch (2,000,000 to 12,000 BCE) on earth was a period of geological history characterized by alternating periods of glaciation and deglaciation. During the glaciation periods, polar ice caps and montane glaciers prevailed and a significant quantity of the Earth's water was sequestered within these ice domains. During the intervening periods of global warming, the polar ice caps and the montane glaciers would melt and recede. Each cycle of freezing and melting lasted for many thousands of years to complete. This geological period of alternating cold and warmth; of ice and water coincided with the tenure the Pleistocene humans.

When the polar ice cap reached higher latitudes of the northern hemisphere (upper Eurasia and North America) advanced; during the periods of maximum glaciation, the ice domain generally extended southward to the latitudinal band that extends along the northern coast of the Mediterranean Sea and into Anatolia in Asia Minor.

Each glacial advance locked up a huge volume of water within continental ice sheets thousands of feet thick, resulting in reductions of ocean-level drops of hundreds of feet over the entire surface of the planet. It also withheld a significant proportion of the water resource that would be available for generating precipitation to feed the surface waters, and thereby changing the configuration of coastlines.

Conversely, as the northern ice domain melted, the denizens of the Africa-Arabia-Asia regions would only experience smaller drylands and a larger waterscape. The rains would, once again, be more plentiful and sure; there would be larger and more reliable bodies of surface water. And, as the temperatures rose and water became more plentiful, plant life would be more robust and the animals would return in greater numbers as well – which meant that there would be more plentiful food for the evolving humans.

Throughout these cold-dry and warm-wet cycles… the Pleistocene humans' bodies and brains continued to grow and develop in response to the challenges of the changing natural environments. The brain not only grew in overall size, but also in complexity and power; it reorganized itself into parts with specialized functions… all of which produced a more powerful and efficient brain-body system.

By approximately 1.8 million years ago, the combined effect of the vagaries of the ice and water worlds would spark the original *raison* for a growing population to begin the first global migration. The first of these planetary colonists took the first steps to enlarge the human domain well beyond the African homeland. By the end of their existence on Earth, they had walked into every corner of Africa, the Eurasian Landmass and the Indo-Australia region.

AND, IN THE PROCESS, THE PLEISTOCENE HUMANS CREATED THE FIRST HUMAN REGION ON EARTH: THE TROPIC OF CANCER REGION...

Chapter 3

The Tropic of Cancer Region

The intermittent glaciations in the higher latitudes produced relatively milder climatic effects in an area of the Earth that is called the *Tropic of Cancer Region*. This refers to a great swath of land that envelopes the Earth in an east-west orientation. It comprises the lands that lie roughly between the latitude 30 North and latitude 30 South, along the Eurasian Mass and Africa.

This region would become the core domain of the Pleistocene colonists. It encircled most of the surface of the Earth which lies roughly between the 30th North parallel and the 30th South parallel. Like the orbital paths of objects in space, the boundaries of the Tropic of Cancer Region also varied due to *perturbances* that derived from changes in the planet's astronomical movements, and the resultant cycles of cold and warmth during the Pleistocene Ice Age.

The most striking thing about the TCR of the Pleistocene Period, was its relative constancy of climate. This part of the globe was insulated from the greater magnitudes of climatic change that occurred in the higher latitudes of the Eurasian landmass. This meant that during the lifetime of an individual or even the oral history of a group, the apparent climate would have seemed generally constant and predictable. There were cyclic changes in climate, but they were not as extreme as in the northern regions. So, the temperature conditions were a favorable constant; but the rains were ruled by the external monsoon phenomena and, therefore, were the main variant that caused Pleistocene humans to constantly move about in search of places with reliable surface water sources.

Chapter 3

Making of the Planetary Colonizer

Throughout the period of human existence on Earth, the Cosmic Mind transmitted coded instructions to certain *Enlightened Minds*. These were sent as intuitive constructs, which often took the form of dreams or flashes of intuition, at various times during human development. Then there are the instructions from the Cosmic Mind, which take the form of coding, which implanted in the neurons of the human brain and the nodes of the human mind. These are like digital computer instructions that provide context and algorithms for carrying out the missions of colonization.

The transmission and reception of these coded instructions may seem to stem from serendipity and seemingly are unrelated to the current human condition, but they invariably have found a practical purpose – as the succession of Enlightened Minds of humanity have developed the necessary receptivity and creativity to apply them to the challenges of colonizing of the Solar System.

So, during the years of early Homo erectus and their evolutionary cousins, the cosmic software applications focused on instructions for developing a superstructure (body) and a controlling entity (brain-sensors) that would create the optimum organism for colonizing the planet Earth, as it existed at the beginning of the Pleistocene Period. Later, during the Holocene Period, the software applications would be transmitted via the mind, as the mature human organism turned to mission of creating a human world – by recreating it as a virtual entity.

The Pleistocene Incubation Period

The 200,000 years that followed the GHAE served as the first *incubation period* for the early Pleistocenes. During the thousands of generations of human development in the incubator region of eastern Africa, the neocortex part of the brain became the command and control center for the development of both the individual organism and the whole genome. This part of the brain would go on to reconfigure itself by creating subordinate functional subregions, to which it assigned specialized functions. The neocortex would thus also become the executive entity which would coordinate the further development of the other parts of the brain... so they could operate interdependently and interactively to carry out the development of the optimum planetary colonizer on Earth.

This development of the neocortex system has been the essence of *vitality* for the human organism; Without it, there could be no sentience; no awareness of itself and of the new complexities of serially-changing environments along the migration paths. However, the ancestral, primitive hindbrain has continued it existential functions of maintaining autonomic operation of the lungs and the heart. Equally important is its function as the natural determiner of the "fight or flight" matrix in the uncertain environment of migration into the unknown, which is an integral aspect of colonization.

In other words, if the hindbrain system is not functioning, there is no life, no awareness or cognizance, and no animation...

The Brain-Body System

Throughout the early phases of the two-million-year *long-count* of human history, the brain concentrated on the reconfiguration of the brain and its cranial vessel, the neurological system, the skin, the internal organs and blood vessels, and the skeletal structure, which provided the powers of transit and manipulation of the external environment to the newly-bipedal humans.

The increasing requirements for processing power in the brain had to be satisfied in some other way besides just growing the size of the cranium, which was limited by skeletal balance considerations and by the size of the birth canal during birth. And, because there also was an upper limit to the number and size of the network of neurons and their linkages, increasing the power of the brain had to be done in other ways. The solution to the problem that was developed by the Pleistocene brain was to micro-miniaturize its neural systems and to maximize the total surface area of the neocortex by folding of its surface. The micro-miniaturization solution also resulted in the continuing reconfiguration of miles of neural wiring and the development of relatively tiny micron-sized circuits consisting of the neurons and their axon connectors.

EVEN AS THE HUMAN BRAIN WAS DEVELOPING ITS POWER OF SENTIENCE AND AWARENESS; LOGICAL THOUGHT AND MEMORY... ANOTHER ENTITY WITH THE POWER TO FORESEE INTO THE FARTHEST REACHES OF SPACE-TIME WAS EMERGING: THE HUMAN MIND...

The Human Mind Emerges

The basic foundations for the design and construction of human mind first emerged during the GHAE. It was originally designed to function as the main receptor for the coded instructions from the Cosmic Mind to Homo erectus during the incubation period in the Rift Valleys of Africa, and as they embarked on their first migrations out of Africa to begin the colonization of the planet Earth.

Then, as the generations of human experience accumulated within the human brain's storage centers, the human mind also became an archive of actionable, useful information, especially in the form of patterns, which would provide insights and *a priori* knowledge. It also would become the common storehouse of human cultural knowledge and experiences, and in which all sensory input and emotions of humankind would be stored... to be accessed by human posterity as needed, to carry out the imperative of colonizing the Solar System.

Thus, the human mind was to become a genomic repository of cellular experience and memories that provided the continuity of wisdom from generation to generation during the 2 million years of planetary colonization of the planet Earth. The human mind would also go on to interact with the Cosmic Cloud of wisdom which could be accessed by later humans.

The Mind of Homo Sapiens

About 300,000 years ago, Homo sapiens emerged, and the human mind began evolving in other ways. Its relationship with the human brain was changing. Instead of being simply a passive receiver of data and information from the brain, the mind now began to actively create new kinds of cognition and learning; the power of intuition for becoming *familiar* with the relevant physical environment.

The human mind now was becoming more than just a passive *database* of human experience. It began evaluating the data and transforming it into useful information; it was applying algorithms of logic in evaluating the data it received; and the human mind was producing new kind of knowledge – intuitive knowledge – that could be used to *see* beyond the horizon of the human brain.

In the millennia that followed, the human mind would be the directing entity that impelled and guided the continuing colonization of planet Earth, and the first phase of the colonization of the rest of the Solar System. This would be seen in the explosion of intuitive and creative thought during the Hellenic Age; the flowering of Muslim thought between 700 and 1500 BCE; the Renaissance throughout Europe; and the age of Positivism, Rationalism and Enlightenment throughout the Old and New Worlds. Most recently, it has been manifested in the great leap forward of science and technology towards the goal of colonizing the entire Solar System.

Walter Gomez

*THE DISCOVERY AND COLONIZATION OF THE
FIRST PLANET IN THE SOLAR SYSTEM...*

Chapter 4

Discovering Planet Earth

The hominids who experienced the GHAE must have felt like they had been transported onto another world. The Common Experience of these individuals had been one of an arboreal paradise; a world in which the trees provided for their life-support needs – literally for the taking. It was also a relatively safe and secure *place,* in which there were no threatening tree-dwelling predators.

This idyllic tree paradise changed completely in the aftermath of the GHAE. The forests were shrinking and the surrounding grasslands were growing… the upshot was that the human precursors had to become ground-dwellers to survive. Now, the natural environment presented the Pleistocene humans with a strange new set of problems to simply survive. Unlike their tree-dwelling hominid ancestors, the most basic existential challenges in their new environment on the ground would require them to make many adaptations.

To begin with, if they were to survive as individuals and as a species, the first humans would have to change the configuration of their bodies, practically from head to toe, in response to the new environmental paradigm. They quickly discovered that their arms and legs, which had earlier formed to enable them to move quickly and almost effortlessly from tree to tree, no longer worked as well on the ground.

At the same time, the first humans were faced with the vital need to improve the power and acuity of their senses in the face of existential challenges posed by the open spaces of the savanna grasslands. The acquisition of water and food now was complicated by the need to avoid becoming food for the other predators in the process. It was the most basic of zero-sum ratios (K/k); where K = Killing and k = killed. Fortunately, as the new human senses combined with more powerful processors in the brain, the Pleistocenes began developing a comparative advantage in the contest for survival of the fittest…

One important consequence of a favorable K/k ratio for humans would be a significant growth in the quantity and in the quality of the calories available for consumption – especially high-value nutrition meat and seafood. This high-protein diet would provide the fuel needed to produce relatively rapid and effective changes in their brain-body anatomy and physiology.

To begin with, the brain needed an increased input of energy, which could only be provided by a greater consumption of calories, especially those provided by a high-protein diet. The brain, in turn, would generate the reasoning power and the hormones that would create intelligent, dauntless and aggressive hunting humans. The cranium, which encased the brain, also underwent a reconfiguration to provide the most efficient space for the growing brain and its visual and auditory sensors. The facial features were rearranged to optimize the distances between each of the sensors (eyes, ears, mouth, etc.) and the corresponding areas of the brain that interacted most directly with these data input centers. The upshot was a progressive rounding of the skull and the flattening of the facial area, and the relocation of the relative position of the eyes, nose and ears… which effectively served to decrease the distance from these sensors to the parts of the brain that interact with them.

The neocortex part of the brain also continued to grow, in size and neural complexity, and began to exert hegemony over the hindbrain and midbrain to ameliorate some of the basic survival instincts of all hominids – such as the fear of fire – which would hinder the prime imperative of developing a "human" that would have the audacity to explore and colonize the planet Earth, first… and eventually the rest of the Solar System. At the same time, the frontal lobe began to use the new algorithms provided by the Cosmic Mind to make more refined changes to both the brain and body of Pleistocene humans.

Adapting To Changing Environments

During the first 1.7 million years of human history following the GHAE, the Pleistocene organism made many changes in response to the environmental challenges of life on the ground; in the savannah country of eastern Africa and, later throughout most of the Earth. One of the most important of these was in the redesign of the superstructure for efficient bipedal locomotion on the surface of the planet – as opposed to a life in the trees. This was vital to the achievement of the immediate objective of creating an organism that could effectively carry out either the decision to both chase down animal prey for food… or run from animal predators.

Another vital adaptation was the development of an internal sensor-brain-body control system to validate and process input data from the external environment, for transmission to actuators within the human system. It was largely this "command and control system" of the human superstructure that would make it possible for Pleistocene humans to migrate – one step at a time – throughout the planet.

In time, the act of walking and running vertically would become *natural* for the Pleistocene humans. It was a balancing act, not unlike learning to walk a tight-rope, and it became progressively easier, as the brain-sensor subsystem and the skeletal system continued to fine-tune their symphony of movement and balance on the surface of the planet.

Walking and running became second-nature, and so did the internal balancing act within the ear, as humans learned to operate in an upright fashion. The eyes and ears also learned how to sense objects at greater distances and to coordinate these sensory images with the weapons that they held in their hands. And, this internal gyroscopic device continued to assist in the maintenance of balance even when moving about quickly, or when throwing a spear while running…

But it was not just a matter of walking and running while throwing effectively. The human brain also had to develop a more effective heart and lung system to capture and distribute oxygen to all the muscles that were involved in running and throwing. At the same time, body hair diminished and sweat glands developed to keep the Pleistocene humans cool, especially during periods of exertion – especially during the hunt for prey for meat.

The upshot of the first 300,000 years of genetic and epigentic adaptations by the Pleistocene humans was the development of a carbon-based entity that would be fully capable of colonizing the Tropic of Cancer Region, and later, the northern regions of the northern hemisphere as well, by 1.8 million years ago.

Thus, the first human *planetary rover* emerged in Africa about 1.8 million years ago; it was, by that time... equipped with exquisite balance to coordinate relatively long legs and sturdy, but flexible feet that were stable platforms, yet also enabled humans to walk or run over long distances as circumstances required... The arms and hands (with their opposing thumbs) were finely-tuned to coordinate with the binocular vision of a sculpturer to create stone tools to enhance the innate power of the system...

Their brain was also redesigned to coordinate and control the activities of the rest of the human system... The command and control system directed the overall movement of the organism and its interaction with the external environments... Power for the human rover was provided by ubiquitous plant-based sources of energy that themselves drew their energy directly from sunlight... And, this human planetary rover could maintain internal temperatures within safe limits by intaking water, and by sweating.

The first "virtual mind" was also being created during this period of Pleistocene colonization, but it would not mature as a separate entity until the beginning of the Holocene Period millions of generations later.

Chapter 5

Separate and Superior

As part of the GHAE experience, the Cosmic Mind had instilled in humans the concept of separateness and superiority, over the Earth and everything on it. This realization marked a seminal moment in the history of the first humans: from that point on, they began their separation from "nature," which they began to see as something separate and over which humans should assume "dominion" in their quest to colonize the Solar System.

The dogma of separateness and transcendence over nature would serve as the rationale and *apologia* for the conduct of humans in the colonization of the Earth and, ultimately, the entire Solar System. It also would become the origin of the principle of separateness and superiority over the physical realm, which would manifest itself in the *utilization* by humans of the organic and inorganic beings around them. This internalized notion gave license to humans for using all organic and non-organic elements on Earth; to use flora and fauna as they saw fit, in pursuit of the Cosmic Imperative to colonize the Earth and ultimately, the rest of the Solar System.

Because of this notion of the superiority and separateness of humans over the Earth; the plants, trees, and animals, and every other aspect of the physical world would be *domesticated* to serve the needs of humans. So, the hides, bones, antlers, and tusks of animals became raw *materials* to make clothes and shelters, as well as tools and artifacts. It also meant that animals could be utilized to do work that was too dangerous, or beyond the physical powers of humans.

The inspired notion of human sovereignty over nature also created the ultimate *alienation* of humans from Earth itself; humans had effectively become *terranauts* on a celestial body, although they were consciously aware of it yet. It was also a kind of declaration of independence from the rest of nature and the blueprint for the utilization of materials.

But there were some advanced beings who would continue to appreciate the oneness of humans with nature and the universe. These were humans who possessed special gifts of perception, intuition and understanding of the physical world around them. They discerned patterns, properties and attributes which enabled them to gain insights into the materials and processes, as well as the universal forces that operated on them.

On a more terrestrial level, many advances occurred as the developing senses of the Pleistocenes empowered them to discern differences in the *feel* of various materials – such as stones and the pieces of wood. Phenomena such as water and fire could now be perceived with greater sensitivity. In short, humans began to discern the true nature of matter: its structure, physiology and properties.

Enter the human mind. It is not a part of the brain; nor is it a physical entity that can be comprehended by the senses... it is more like a *hologram* that was created and transmitted to humans via pure energy, rather than through genetic fluxes, and which has continued to develop in concert with the human brain.

The intuitive powers of the mind can access the secrets of the universe to present the possible ways to overcome the obstacles that appear in the quest to colonize the solar system. The human mind derives the input of light and sound, in conjunction with stored *Cosmic Memories,* to construct a shared *Cosmic Reality...* which is the repository for summary experiences of humans – over time and from place to place.

Chapter 6

Ways and Means of Colonization

Realizing the great variety of environmental challenges that would be faced by humans as they attempted to colonize a dynamic solar system, the Cosmic Mind transmitted to Enlightened Humans the insight to guide both organic and cultural adaptation to changing environments. This has been manifested in the serial advances in the "knowledge-matrix"; the "tool-kit"; the "skill-set"; and the "cosmic-consciousness" that emerged at every point in the human quest to colonize the planet Earth.

Another overriding theme of human colonization has been the development of artificial augmentation of the innate powers of the organism – by developing knowledge; tools; skill-sets; and the social organization to carry out the Cosmic Imperative. So, within a relatively short time – about 200,000 years following their emergence on Earth – the early Pleistocene humans already had advanced beyond the point of merely using the sticks or stones that happened to be lying around them as tools or weapons. Instead, there was an emergence of conscious forethought, planning and learning to convert natural objects into manufactured tools and weapons to be used for specific purposes. And, perhaps most importantly, humans were using the power of communication and cooperation to give greater leverage to the tools and skills they were developing.

So, throughout the course of the Pleistocene colonizations, the processes of developing the suite of knowledge-bases, tool-kits and skill-sets continued to become more sophisticated and elaborate: data points and specialized skills were mimicked and otherwise propagated through advances in oral and symbolic communication; tools and weapons were becoming more efficacious. Thus, long sticks became aerodynamic spears and rocks were fashioned into refined hand-axes. Later, the spearheads gradually got sharper and more specific in their designed function of killing prey. More generally, the conversion of natural objects into weapons of the hunt can be seen as a metaphor for the general separation of humans from the natural to the artificial, by following the same trajectory of transmutation in which tools and skills become increasingly sophisticated in achieving purpose.

Also during this phase in human history, the basic human model for technological advancement continued to develop. One characteristic of this model was the distinctive *iterative* approach of persistent testing and retesting in the development of new technologies, in the face of changing local environments and situations.

Perhaps the most powerful force in the colonizing of unknown environments has been the strategy of using small cells of committed humans, working as a determined team, to carry out long-term objectives, and to propagate these cells throughout a territory by the process of replication. This explains how a relatively small population of Pleistocenes could – with little more than their own brain-body system – ultimately *domesticate* the natural environments of the Tropic of Cancer Region in less than two million years. It also explains how the Holocenes could restart the colonization of Earth, even after the cataclysmic events of the end times of the Pleistocene Period, and how the Solar System would be colonized within one century, during the Holocene *times of troubles* of the 21st century.

The deployment of small groups of humans, each optimally designed and equipped to operate at a grass root level, has proven to be a highly effective organizational construct for carrying out planetary exploration and colonization. It also has been an essential element of a deliberate strategy for propagating the human genetic material throughout the varying environments that are colonized.

PLEISTOCENE HUMANS BEGAN TO UNDERSTAND THE SECRETS OF NATURAL FORCES AFTER THE GREAT HUMAN AWAKENING EVENT...

The development of appropriate adaptations to the effects cataclysmic forces of fire and water; earthquakes and volcanic eruptions; and of boulders and sands were paramount to the human colonization efforts on Earth.

Consider that the growing neocortex enabled the first planetary colonizers to overcome the primitive part of their brain's natural fear of fire. Thus, by using their enhanced powers of cognition and rationalization, humans became the first animals to overcome their natural fear of fire and learn to use its properties to ward off predators. This familiarization with fire then enabled them to utilize the heat of fire for warmth against the cold and to ward off the predators of the night. Equally important, was the opportunity it provided for relaxing and doing the work of socializing and strengthening the bonds of unity for the members of the group.

Pleistocenes also developed an imitate *familiarity* with the force of gravity: to stand erectly and to effectively negotiate with their gravitational environment as they used their bodies to carry out the tasks of surviving and thriving on the Earth. Consider, for example, how these early humans learned how to extend the length of their arms and twist their torso to hurl objects over greater distances, and thus overcome the drag of gravity by applying more lateral force to the spear. A tool they devised for augmenting their natural powers in such a maneuver was the "throwing stick." With it, they could multiply the force and distance of their hurled spears. They also discovered the secret of stored energy when they developed the curved wood bow to propel arrow projectiles.

The Pleistocenes gained at least a "working understanding" of gravity on the surface of the Earth. They developed a practical understanding of how to manage the natural tendency of planetary gravity to resist efforts to move or lift heavy objects. They therefore understood, at a practical level, that they had to exert a given level of muscular force to propel their spears… depending on their length, girth and, therefore weight. They also experienced this increase in the difficulty factor when they threw rocks and clubs of greater sizes.

They also learned the concept of *negotiation* with gravity. So, at some point, they realized that the combined lifting, pushing and pulling efforts of several persons could exert force on more massive objects, much more efficiently and effectively than one set of arms and legs; that human power could be multiplied with augmenters, such as ropes and poles, to *leverage* the human organism's ability to manage the gravitational attraction of the planet at a local point on the surface of the planet.

SO, UNDER THE DIRECTION OF THE COSMIC MIND, HUMANS INITIATED THE FIRST PHASE OF THE COLONIZATION OF THE SOLAR SYSTEM...

Chapter 7

The First Planetary Colonization

During the 200,000-year period of incubation in Africa, the Pleistocene humans had developed the necessary brain-body architecture and systems for colonizing the planet Earth: one place at a time, by one small group at a time.

So, under the direction of the developing neocortex and the programmed instructions received from the Cosmic Mind, humans initiated the first phase of the colonization of the Solar System, when they began colonizing planet Earth.

The Pleistocene colonizers were autonomous "planetary rovers." One can imagine each clan of these earth colonists as being like the modern robotic rovers that explore the surface of the other planets of the Solar System, except that the *terranauts* didn't need to carry artificial oxygen or make artificial gravity to survive; nor did they have to manufacture an artificial "earth-like atmosphere" within enclosed space structures. Every element of life was already present wherever they ventured on their home planet. And, unlike on most other places in the solar system, earth's lower atmosphere (below about 20,000 feet of altitude), provides the optimum mix of oxygen and other gasses, as well as the various states of water that are essential for human life. This natural hospitality is also fostered by the outer envelopes of magnetic fields, ozone layers, and fields of charged particles that protect the terranauts from harmful cosmic and solar radiation.

Manifest Destiny

In the aftermath of the GHAE, humans were imbued with a powerful drive to explore and colonize the entire Solar System—beginning with the planet Earth. The powerful will to colonize became part of the natural and cultural genetic code that was propagated throughout the developing genome.

So, about 1.8 million years ago, the first waves of human colonization of the planet Earth began. These primeval *terranauts* had already developed the optimal suite of sensors with which they could detect and transmit to the human brain the essential environmental data that would be needed to colonize the planet Earth. Thus, during the incubation period in Africa, their eyes had already evolved the necessary depth-perception and high-resolution to operate effectively and efficiently on the surface of the planet. Their visual sensors could also automatically refocus from long-range to short-range, which provided them with a refined system for gauging the distance and movements of phenomena.

Another set of sensors provided auditory data to complement what they were seeing. These too evolved bilaterally to provide a 360-degree experience of external sounds. It was the stereo feature of the visual and hearing systems enabled them to ascertain the distance of the source of the sounds with greater precision. And, a more sophisticated network of sensors in the skin provided another flow of continuous data for processing by the brain, thereby providing human colonists with the level of awareness necessary to adapt to challenges or opportunities in any physical environment.

Equally important was the development of *cultural memories* which were carried within each cell of humans. It is this genetic archive of the sum of experiences that becomes part of the human genome. Indeed, each cell is itself a powerful repository of knowledge that can be used as a resource by the species, in response to changing environmental challenges and opportunities. Such diversity of genetic materials would prove to be vital in the continuing colonization of the home planet.

Geospace 2060

A UNIVERSAL STRATEGY OF PLANETARY COLONIZATION IS EXPRESSED BY THE DICTUM TO "FOLLOW THE WATER" ...

The Cosmic Mind taught the humans that they should always *follow the water* wherever they migrated and settled. On Earth, this meant that they should follow a strategy of migration and settlement along the littorals of oceans, gulfs, rivers, and lakes. The shores and banks of these bodies of water offer the path of least gravitational resistance to movement and to lifting and carrying of objects. They also offer plenty of highly-nutritious seafood and an abundance of plants to forage and animals to hunt on nearby lands. And, along the edges of rivers and fresh water lakes, there is water to drink.

During the first 1.5 million years that followed the GHAE, the planet Earth seemed to be a place of overall constancy – at least as far as each generation of human colonists could discern. Wherever the Pleistocene colonists went – within short-count of generational memory and within the confines of the Tropic of Cancer Region – they were almost always presented with warm-wet or cool-dry weather conditions. In either of these climatic regimes, the strategy of following the littorals of bodies of water was an optimum strategy for getting sufficient water and food to survive and thrive; to disperse and become diverse.

Geospace 2060

*THE PLANET EARTH UNDERGOES A
CATASTROPHIC CHANGE...*

It began about 300,000 years ago… the relationship between the waters and the drylands on the surface of the Earth began experiencing major global changes. The polar ice sheets and the alpine glaciers melted and released waters that invaded the drylands. In the process, familiar coastal lands and river beds, which had been the most desirable for settlement, disappeared and caused major relocation of large populations of Pleistocene humans. It also stymied any advances in developing more settled ways of life for many generations.

The long-count effects of global changes were manifested by short-count periods of seeming constancy within the life-experience of an individual or clan. This meant that each generation of human colonists continued to experience what seemed like *normal* weather conditions; an encapsulated set of environmental challenges, which could be dealt with by short-count cultural adaptations.

Meanwhile, in the long-count experience of global warming, there was a continuing, gradual subsuming of land bridges, which eventually created gulfs of separation that blocked the movement of colonists between previously connected landmasses. And, eventually, landmasses eroded, thus creating archipelagos and islands. All these tended to produce isolated pools of genetic materials, especially in the southern hemisphere.

But, to any given group of Pleistocenes, the presence or absence of a land bridge would hardly have been noticed: the existing landscape, the bodies of water and the horizon were simply *there*, and the migration would continue… or stop at the water's edge. But overall, the disappearance of the intercontinental land bridges would pose a near-existential threat to the survival of humanity on the home planet, as migrations encountered bottlenecks and some populations were isolated. Yet, wherever land-bridges survived, some ripples of migration and cross-migration persisted, until the land bridges were finally subsumed by the waters. It was only this happenstance that enabled the human species to survive the genetic bottlenecks that nearly caused its extinction on Earth.

Thus, throughout the period of global warming, which occurred in the Pleistocene and Early Holocene periods, the colonists continued to follow the natural paths provided by the shorelines of large bodies of water that rimmed the landmasses. These were places that required the least energy to traverse, they were filled with food resources rich in protein and other important nutrients, and were adjacent to lands that offered other food and water to sustain the migrants.

So, it was the lands that border the Red Sea were one of the first such natural pathways that were utilized by the human colonists. These led them naturally to the Levant and, later, to the littorals of the eastern Mediterranean Sea and those of the Black Sea. Also, as the migrating colonists followed the contours of the Red Sea coastlines, they would come to a point where they could cross over onto the Arabian Peninsula… then follow the contours of the Persian Gulf onto southwest Asia… all the way south to the islands of Indonesia… and ultimately, the southernmost landmasses of Australia and Borneo.

One way of comprehending the success of the colonization of the planet Earth by the Pleistocenes is by mapping the spatial distribution of the hand-axe at archeological sites. This ubiquitous hand tool emerged, in its simplest form, during the earliest stages of the incubation period (approximately 2.5 million years BCE). It was introduced by Pleistocenes in Africa, and it was then propagated throughout much of the Tropic of Cancer Region, as well as outlying territories in Australasia.

The Levantine Rendezvous Region

The Levantine corridor is a relatively narrow strip of land, which connects Africa with the Arabian Peninsula to the southeast, and Eurasia to the north and east. Pleistocene humans first used it as a "land corridor" between Africa and Eurasia.

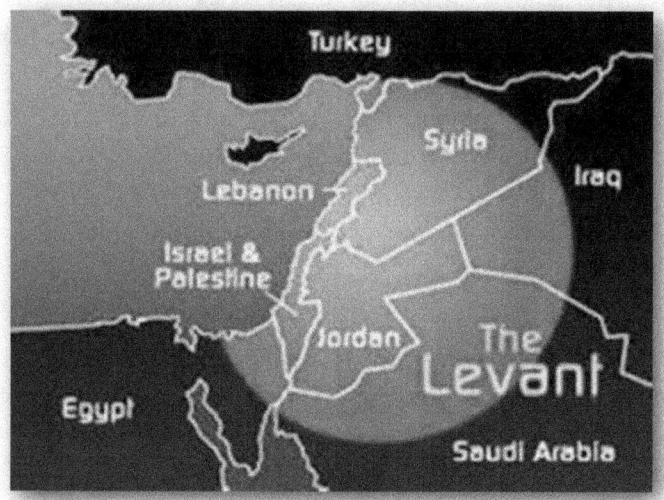

THE LEVANTINE CORRIDOR

The Levantine Region was a natural trajectory for ancient humans in the colonization of Earth, primarily because of its proximity to the eastern shores of the Mediterranean Sea, which then provided a coastal route to the littorals of the Black Sea. Even during the earliest phase of Pleistocene colonization, the Levantine Region already was a central hub for continuing waves of migrations of populations throughout the Tropic of Cancer Region.

The Levantine Corridor also provided easy access into the northernmost frontier of the natural comfort zone of Pleistocene humans, especially during the glaciation periods, when the passage was exposed by the retreating waters. There, in the northern frontier of the Tropic of Cancer Region, it was somewhat colder than in the Rift Valley homelands.

But, with their rapidly-developing neocortex, the colonists continued to improve and adapt their tool-kits and skill-sets to meet the new challenges.

Thus, they added new features to the basic hand-axe: giving it sharper edges by more nuanced working of the stone; and making it ergonomic and easier to hold. The basic hand-axe also became a precursor to the smaller stones that were affixed to a shorter version of the spear and many other implements and artifacts.

In addition to creating a greater repertoire of tools, the Pleistocenes also were constantly learning more ways to employ the properties of fire: to use the light of the campfire and the torch to extend the work day; to make the processing of game animals more efficient and to make raw meat and produce tastier and more nourishing. They also became more adept at using virtually every part of the animal carcass: to tailor the furs and skins to make clothing; to make tools and weapons out of the antlers and horns; to fashion bowls and other containers; and they learned to use both natural and artificial materials to construct better shelters.

The upshot is that the first wave of migrations would result in the development of permanent colonies as far north as the Levant Region and Asia Minor, as early as 1.8 million years BP, and only about 200,000 years following GHAE event in Africa.

[This would prove to be greatly significant to the success survival and propagation of the human species on Earth; the Levant would become the place of existential refuge for the Pleistocenes and the birthplace of the new Holocene colonists.]

And, during this time, there were already developing proto-types of divisions of labor and spaces within their settlements. In a remarkably short period of evolutionary time (a few thousand generations, perhaps) these early humans had appreciated the concept of division of labor and production. So, already there was a division of labor and specialization, in which the hunters and gatherers of food and other raw materials and the processors of raw materials developed into separate sectors of the nascent economy.

And, the "hunting-grounds" would soon be recognized as natural regions, with their own boundaries and internal properties and characteristics. There would also emerge a consciousness of "comparative advantage" – which is the notion that some hunting-grounds or places within it have special advantages with respect to other places in the area; and with respect to given criteria. Usually this meant that a more reliable source of water was nearby, which meant that there would be plants, which would attract animals and, therefore, there would be foods resources for humans. In some cases, there would also be fish and other aquatic animals that could be used as food sources and as materials for manufacturing artifacts.

There also were certain places that were nearer to useful stones and wood for making tools. It would be there, in those places with comparative advantages over other places in the hunting ground, that some humans would abandon the hunt and instead dedicate their time and energy to making clothing, tools, and other artifacts from the carcasses of the animals that others had hunted. This would be the original formal division of labor and bartering in the Pleistocene Period.

Chapter 8

The Pleistocene Human Region

As indicated in an earlier narrative, The Tropic of Cancer Region (TCR) was relatively less affected by the Pleistocene advances and retreats of the polar ice caps, than the more higher latitudes of the northern hemisphere. Within the TCR, the temperature regimes and the bodies of surface water varied little during the lifetime of a planetary colonist. And, even during the long-count of several human generations, the changes in temperature regimes were less significant and the waterscape was either reliably wet or expectedly dry.

And, during the short-count generations of human colonists, the familiar land bridges and coastlands remained faithful to their community memories – until they didn't. However, from the perspective of long-count periods of thousands of generations, ancient land-water configurations would change inexorably, until familiar landscapes and waterscapes finally disappeared from generational memories. And, finally, the distant polar ice sheets and montane glaciers began an irreversible melting, and began to release their cache of waters into the atmosphere and onto the depressions of the planet's surface.

Within the Tropic of Cancer Region, these liberated waters began to encroach on the perimeters of the dry lands in a series of seasonal waves, but almost imperceptibly within the short-count memory of any human generation. Nevertheless, over the long-count of several generations, some familiar lakes and some previously reliable rivers would diminish or disappear; and there would come a time when dry *bridges of land* that gave human colonists free transit from one large landmass to another would no longer exist; the waterworld would ultimately erase their existence.

[However, an exterrestrial observer of Earth would still have perceived the Tropic of Cancer Region as a series of peninsulas and archipelagos extending southward from the main Eurasian landmass, and plunging into great expanses of oceans. They would have noted that they might have been formed by the tectonic separation of an erstwhile single landmass.]

Within the area encompassed by the Tropic of Cancer Region, the cyclic changes in the ratio between the water world and the drylands also created generational weather regimes that were either hot and dry, or hot and humid. The local and present vegetation was cyclically and variously described as a rainforest-savanna mix; as a combination of low bushes and grass clumps; as a tropical evergreen forest; or as a tropical, semi-deciduous forest.

Any of these local environments would have reminded a given generation of Pleistocene colonists of the ecology of their erstwhile homelands in the Rift Valley of Africa. At the same time, the strategy of remaining close to the water in any ecological situation would have been very effective in providing the Pleistocene colonists with a generationally familiar set of natural environmental conditions wherever they ventured, regardless of the global water-land ratio.

THE FIRST WAVE OF HUMAN COLONIZATION OF EARTH ENDED WITH THE PLANETARY "TIME OF TROUBLES"

Chapter 9

The Holocene Colonization

These were the Pleistocene "end days," but they were also the "beginning days" of a new generation of humans – the Holocenes. They were the product of the generations of migration and cross-migration, which was precipitated by the terminal global warming and flooding event that ushered in the Holocene Period of human history. So, about 30,000 years ago, a new generation of humans became the new planetary colonists on Earth. They began with a few colonies, which they located at *favored places,* where water and food resources, and the other essential elements of human survival happened to coincide.

In the beginning, many Holocenes still followed a nomadic, hunting-gathering way of life. But gradually, over many generations, some of them decided to settle permanently along the banks of rivers or at oases where natural springs were found, or along the shores of inland lakes and the coasts of the oceans. Several populations of Holocenes would be established in what is now China; others in Pakistan, Mesopotamia, along the eastern and southern littorals of the Mediterranean; and still other clusters of colonies would be established around the Black Sea.

The colonization of the home planet by the Pleistocene humans had followed a strategy of constant, purposeful movement, a form of nomadism – from one hunting place and resource site to another – in their search for the basics for individual and group survival, and the propagation of the human colonies throughout the Tropic of Cancer Region. As has been discussed in earlier narratives, this strategy proved to be optimum given the prevailing environmental conditions. However, when the surface of the Earth was transformed at the end of the last Ice Age, a new variant of humans, the Holocenes, decided on a more proactive, controlling strategy for survival and for re-colonizing the planet.

This would be a strategy that was based on permanent central places near sources of water, surely for drinking, but also to irrigate the plants that were being domesticated to provide nutritional cereals on a large-scale. Water itself would be brought under the control of humans; it would be subjected to engineering in the same way that rocks and sticks had been fashioned into useful artifacts and tools.

The Holocene approach to surviving and thriving on the renewed Earth began with an inspiration, a vision of some Enlightened Minds, which would impel the Holocene colonists to perceive a given area as a definitive space, with a boundary and in inner area that could be "domesticated" the same way as fire had been. Thus, a definite plot of land would be brought under human control and utilized for specific purposes, such as a for producing food and fiber. These were areas to be organized and utilized as places to cultivate plants, rather than simply harvesting them in the wild. These eventually became places to husband, rather than hunting, animals for meat and hides.

These became spaces which would be measured and surveyed; to be subdivided into functional sites, for living and for working; and there would be common spaces for economic and social intercourse. Gradually, the concept of a space as a "field" entered the consciousness of these ancient planetary colonists. The "field" [like the field of the physicist] as a virtual space in which human activities had to follow the "rules" of the environment became a reality. To the early Holocene humans, the field would take the form of a tract of land; a definite area, having a recognized border or frontier which define the limits of the field. It likely would be manifested by fences to protect the "insider" resources and people from the "outsiders" – especially the two and four-legged kind that might seek to take away the foods they had "created" by the work of their brain and body.

The new strategy of "staying in place" was also characterized by a spatial system of organization for producing varieties of food, fiber and other goods. Planting-grounds would be organized, subdivided and managed; labor would be organized and managed; non-farming economic activities would be organized and managed – and the whole economic system would be managed by various accountants, scribes, and other overseers.

At the same time, the Holocene planetary colonists also began to lay the conceptual framework for creating permanent *places* as sites for creating societies with more definitive shared cultures and, ultimately, *civilizations*. These became central places which, because of their comparative advantages (natural and cultural), usually evolved into advantaged rendezvous locations, where the local roving bands of hunter-gatherers would be drawn to gather; where some folks would settle into a new, more permanent way of life. For the most part, these were places that were situated near sources of water, fertile soils, and necessary natural materials. There also were special sites that would serve as logistical centers, where concentrations of natural resources (such as special rocks or woods) could be *cached* or warehoused for use by the local rovers or transient colonists.

INTENSIVE FARMING ON STATIONARY UNITS OF LAND REQUIRED THE INTENSIVE INPUT OF MANY HOLOCENE HUMANS...

The Holocene colonists began to nurture, cultivate and harvest certain plants which could produce large quantities of calories and nutrition, in the form of cereals, which could be more easily quantified and distributed, and the surplus could be stored as needed. These were desirable food plants, to which water could be directed at the appropriate time and place to maximize the quality and quantity of food output.

Generations later, agriculture developed into a sophisticated system of plant cultivation, which required a group effort at every stage of the food-production process. Whether preparing the fields for planting; developing the irrigation system; gathering and planting the seeds – that is, dealing with all the variables of agriculture – teams of people, with various specialized knowledge and skills were needed to be successful in the endeavor.

And, because virtually everyone was actively involved in the food-production industry, a common frame of reference developed. Many brains, eyes and hands were dedicated to a common activity, which required coordination of efforts, and soon there began to develop a system of oral communication and body language that facilitated the coordinated efforts of the group. In time, there emerged a more complex system of communication, which included a combination of vocalizations, facial expressions, and the "language of the hands." All these forms of communication were employed to exchange ideas and notions related to farming and animal husbandry. And, from these commonalities of experiences, written symbols would become literature and vocalizations would be codified to produce language and literature.

Empathy and Communication

The mechanism that was developed by the neo-cortex for communication among Holocenes was based on a highly-developed power of *empathy*. Putting one's self within the thoughts and emotions of another human became another great leap forward in the process of communicating; this is what enabled them to transmit their thoughts via scratchings in the dirt; an assemblage of stones or sticks; the simplest of oral soundings or clicks; and the most powerful form of interpersonal communication – the facial expression.

Holocene humans also learned to communicate and learn by observation of another's actions and body language, and then developing *mirror neurons* in their own brains. It was a physiological mechanism for converting observation into learned behavior that would be key to managing the growing universe of human activities. With the power of *mirrored communication*, humans would go on to colonize the entire TCR and the Eurasian landmass.

THE HOLOCENE COLONISTS WERE ALSO DEVELOPING THE FIRST HUMAN SPATIAL SYSTEMS...

Chapter 10

Human Spatial Systems

Nodes and the linkages that connect them are the basic components of all spatial systems. Human places and human interactions are the nodes and linkages of human systems (regions). From these basic propositions, It can be logically asserted that all human activity occurs within a given space. Empirical observation and logic also lead to a conclusion that a space can be small or large along several dimensions. These spaces also contain physical, biotic, and cultural phenomena. And, when these spaces can be surveyed, organized and developed for human purposes, they become human regions.

Humans have always occupied a given *place*, at a given time; the process of moving from place to place created a *space-time* link between events; and different space-time events were linked by visible or invisible conduits. This has been this essential dynamic has continued to define the colonization of Earth and of the rest of the Solar System.

It is no wonder then, that, within 200,000 years following the GHAE, the Pleistocene humans had already developed a well-defined and operating *spatial system* (or cultural region) of nodes and linkages in the Rift Valley of Africa. About a million years later, the Holocene humans would follow the same basic blueprint in developing a somewhat different society within the Tropic of Cancer Region. In both cases, the planetary colonists established a spatial system of interconnected and interdependent nodes.

The universal strategy that all humans adopted in colonizing the home planet can be characterized as one of dispersal and diversity. In practice, it was based on widespread dispersal of nodal settlements... connected by the physical movements of humans and their *payloads* (payloads include the human and containers, goods, ideas and so forth). During this first period of planetary colonization, each human was effectively a *planetary rover* and, more abstractly, a carrier of photons of energy, atoms of elements, and cells of genetic material. Thus, the individual migrant or traveler represented the essence of the inter-nodal links that animated and energized the Tropic of Cancer Region.

Human colonists on Earth have always been rovers; both as hunter-gatherers and as intensive farmers – relentlessly in search of water and food first, and then the growing list of other organic and non-organic natural resources – to survive and prosper as individuals and as societies. Even in the Holocene time of stationary, intensive resource development, the restless urge to move from place to place proved to be a highly successful mechanism for maintaining the vigor of the human species on Earth, and a dispersed and diverse pool of genetic materials provided the early humans with the best chance for achieving the ultimate objective of colonizing the entire Solar System.

So, if a flood or drought; or an earthquake, seismic event, or tsunami... should occur in one region of the growing Tropic of Cancer Region, the human survivors – given their mobility and survival skills – could easily migrate to create new nodes or to assimilate into existing ones. Again, the strategy of dispersal and diversity fostered the survival of the resilient genetic pool that enabled the human species to survive every existential threat to the species.

At another level, the strategy of dispersed propagation and the development of a hierarchy of central places enabled Holocenes to colonize a great swath of the planet earth over the course of about a million years. They did it by maintaining roving groups of resource hunters and pathfinders who inexorably surveyed the surface of the planet for useful minerals and other natural resources that could be used to sustain the stationary places of the Holocene system.

Even before the start of the Holocene Period, some humans began establishing places where the product of a hunt or foraging could be processed and consumed. At first these were simple, transient camps, but some develop into more permanent settlements. The main reason why certain places became *Central Places* was *comparative advantage.* That is, they were situated close to water and therefore, they were places where plants and the animals congregated.

Sometimes these centers of food resources were also situated near sources of certain rocks and trees which offered materials for making tools and constructing shelters. These centers of raw materials often were sufficiently advantaged with resources that some colonists could afford to specialize in the extraction and processing of raw materials to manufacture finished goods sale to the food-producers, other specialists and transients. And, sometimes the specialists would produce a surplus of finished goods – which they could trade for exotic materials and goods.

As the number of Central Places grew and became more diversified, they became a vital component of the growing spatial system that was first developed by the Pleistocene humans and later refined by them. Like the neurons of the brain; they were places where human travelers, goods, technological information, and social wisdom were stored and exchanged.

It was during the Holocene colonization that some Central Places also provided a significant *surplus* of food resources; more food than the local denizens needed for their own subsistence. This meant that some the non-farm sector of a colony could devote its energy to the production of more labor-intensive manufactured goods, not only for local consumption but to trade with other colonies as well. In time, the earliest manufacturers would grow their special knowledge, and hone their skills and tools. Some became masters in stonework; others advanced the utilization of wood and fiber. Others would specialize in the processing of minerals and metals.

Perhaps most importantly, a social contract developed between the individual and the society of a colony. So, within 200,000 years of the GHAE event, Pleistocene humans were already introducing many of the social constructs that were to become the norms of social interaction, which guided how individuals would congregate to exchange, not just goods and services, but also ideas and mirrored neuron knowledge and skills.

Many Holocene central places became specialized *learning centers* where resident shamans, muses, priests and other enlightened beings dispensed wisdom and justice to supplicants who sought them out. These oracles were humans who had achieved the highest level of wisdom derived from mirrored eons and the highest level of neural systems…

Chapter 11

The Making of the Holocene Planet

As noted above, the surface of the planet that the Pleistocene colonists had known for more than a million years began to change drastically about 300,000 years BCE. The process began with some long-count perturbations in the orbit of the Earth around the Sun; which instigated the beginning of a long-count warming of planet earth, due to significant changes in the gravitational relationships between the Sun, Earth and the Moon. This rearrangement of Earth's spatial relationship with the Sun would be manifested in prolonged, persistent warming of the globe, global flooding, and major tectonic changes on the surface and within the upper level of the mantle.

First, the planet began to experience a series of longer than usual short-cycles of gradual warming that set into motion a relentless melting and diminishment of the ice caps in the higher latitudes of the globe and in glaciers in high elevations of the surface. Polar ice caps melted, glaciers receded and both continued to empty their now-liquid waters into the oceans, seas, and inland depressions of the lithosphere, thus creating many large and small lakes. The flowing waters sought the path of least resistance and thus created a myriad of streams and rivers. In the mountains, glaciers also continued to melt and their waters also fed into the developing systems of rivers and lakes.

Next, as the waters on the surface of the planet were being rearranged, so too was the great weight of the water-mass being redistributed: from the poles to the lower latitudes and from the higher altitudes to the deepest depressions of the Earth's skin. The entire surface of the planet felt the dynamic effects of the shifting weight of the waters. This shifting of mass on the Earth's surface also produced many effects within the mantle of the planet. The constant flexing of the surface produced friction heat which radiated throughout the outer layers of the planet, thus causing a heightening of the convection of material between the surface and the interior of it.

On the surface, this inner turmoil of the planet would be manifested in an intensification of seismic and volcanic events that would punctuate acutely the long-count remaking of the topography and the surface hydrology of the Pleistocene world – which was the only habitat that the human colonists had ever known. So, even as the they were driven from place to place, as refugees from one natural disaster after another and – more importantly, from their ancestral hunting-gathering sites, by the earthquakes and volcanic eruptions, large-scale conflagrations and mega-tsunamis which plagued them for thousands of generations, the Pleistocene humans persisted in following the old ways of life.

Meanwhile, huge masses of ice continued to be transferred to massive bodies of liquid water, adding to the depth and size of the oceans and seas of the planet. The relentless encroachment of the waters along the coasts of the landmasses often created ragged boundaries between the water and dryland worlds, thus extending the total length of the shorelines, and creating semi-enclosed cul de sacs of relatively placid waters of the inlets and bays.

Such global rearrangements of the waters had occurred earlier in the history of the planet, but this event was different: humans now occupied the drowning coastal areas. This time, the global encroachment of the oceans, seas, lakes, and rivers on the adjacent drylands drowned water-side human colonies; and the subsidence of the intercontinental land bridges directly disrupted the normal movement of humans as they migrated to form new colonies or assimilate into existing ones.

The land bridges which the terranauts traditionally used disappeared within the span of two or three generations as they began disappearing beneath the waters. This meant that the Pleistocene *terranauts* could no longer simply walk from one end of the Tropic of Cancer Region to the other. The rising sea levels – hundreds of feet – were redrawing the map of the planet within the life span of present generations. Some coastal areas were nearly or totally engulfed, creating many new islands and archipelagos and stranding their human populations.

To an extraterrestrial observer, orbiting at an altitude of 200 miles above Earth, the overall reconfiguration of the surface waters relative to the drylands would have been the most significant change on the planet. They would have noted that this phenomenon was the basic driving factor in collateral changes within the atmosphere and even in the magnetosphere.

In fact, this was the beginning of the *end times* for the Pleistocenes. They would be harried by slow-motion tsunamis that disrupted the cultures that had been created along the shores of the Indian Ocean, the Black Sea, and the Mediterranean Sea. Some Pleistocene cultures virtually were wiped out, as in the case of landmass north of Australia which disappeared like the fabled Atlantis.

They had already established colonies on most of the surface of the planet. They had reached beyond the northern and southern frontiers of the Tropic of Cancer Region… as far as northeast China and into the southern landmass of Australia. Only the Americas and the polar regions remained unexplored. And, they had accomplished this feat by simply walking, one step at a time…

And when the planet began to change, it was barely experienced by a given generation of colonists. As far as each human was concerned, the world about them was what it was; the local and temporal space and time seemed to present a normal set of existential challenges. But to an external observer with a long-count perspective, the global changes that were occurring were nothing less than cataclysmic and understandably, they impelled serial generations of Pleistocene colonists to migrate, migrate, migrate… back to the ancient heartlands around the lands of the Mediterranean Sea and the Black Sea, and along the beds of the Nile, Tigris-Euphrates.

So, when the traditional lowland habitats finally disappeared beneath the rising waters, many populations of humans now, for the first time in living memories, found themselves restricted to smaller land areas, and effectively cut off from other isolated clusters of humans throughout the erstwhile global domain of the Pleistocenes. This was a time of existential angst for the human species.

The global encroachment of the waters onto the drylands had produced a new surface-scape of diminished continental landmasses, drowned intercontinental land bridges, and newly-created islands, usually the tops of erstwhile mountains or volcanoes. The overall rise in the magnitude of the global water domain also effectively redrew the map of possible migration routes for humans: it created a new system of pathways for migrants; it also produced roadblocks and other perturbations in the movement of refugees and colonists across the surface of the planet.

In their wake, the rising waters rearranged the spatial distribution of the Pleistocene societies. They became more fragmented and, for hundreds and even thousands of generations, the survivors of this existential challenge sought to create new nodes and linkages to form a new system of colonies. Equally significant, these changing islands of DNA and culture were producing a multitude of genetic mutations and cultural variations that would later emerge and recombine in the Holocene human populations of the Mediterranean Sea and Black Sea Fertile Crescent Region and the Black Sea domains.

Amidst the turmoil of the end times of the Pleistocene Period, the food resource base of the planetary colonists also underwent significant changes. The animals which had been hunted for meat and fiber experienced the same existential challenges as the humans; and many would lose the game of survival. The remaking of the planet was one reason for the extinctions; another was that the late Pleistocenes had become extremely proficient at hunting prey. In fact, they did so well that they serially caused the first human-originated extinction of certain animal species within their local *hunting regions.*

Because of the extreme changes in the environment and ecological systems, the Late Pleistocene colonists could no longer depend on hunting and gathering as a viable survival strategy. Nor could they survive as a species by resorting to a nomadic style of farming, which is referred to in modern times as *slash and burn* agriculture. They would have to find new systems and processes; they would have to *terraform* the surface of the planet if they were to survive as a species – especially if they were to successfully prepare for the ultimate colonization of the rest of the solar system. Their only option was to settle down, to "develop" the local land, to create an artificial food production system – one in which they could minimize the risk variables and maximize the caloric output from such a system.

Chapter 12

The Holocene Human Incubator

At the dawn of the Holocene Period, a new generation of humans would develop new incubator regions on the planet. Thus, by about 30,000 years ago, the survivors of the first attempt at human colonization of a planet had migrated back to the littorals of the Mediterranean Sea and the Black Sea in the north. Other refugees of the planetary makeover had resettled in northeast Asia and in the peninsulas of southwest Asia. Most were modern Homo sapiens, some were Archaic Humans, and a few were remnants of the Homo erectus humans. Since that time of renewal, these various human populations have interacted and intermixed to produce a new Holocene Human.

These centers of renewal became the places where a new a new strategy for producing food and fiber developed; one based on intensive food-production systems. These would replace the old nomadic, hunting-gathering ways of life; they would result in new ways of settling and colonizing the renovated surface of a planet; which would ultimately create a second incubation period for the human species on Earth.

The first serious attempts at developing domesticated food production systems began around 20,000 years BCE. By then, the reconstruction of the Earth's surface and the rearrangement of the land-water ratio on the surface of the planet had almost ended. And with the sense of some basic degree of stasis and stability, the plants and animals that had survived the millennia of *environmental troubles* resumed their adaptations, within the context of the new environmental paradigm.

The new human centers of the Tropic of Cancer Region developed where there were reliable sources of water. They also were blessed by a reliable incidence of solar energy and with soils that were constantly reinvigorated by predictable cycles of river sediment. The myriad of permanent rivers, which drained into several inland lakes, also created new opportunities for developing aquatic food production systems; providing opportunities that could be exploited much more intensively by the settled Holocene peoples.

Nevertheless, the new Holocene colonists would still have to continue to make many adaptations: to continue to develop their brains and neurological systems; their powers of cognition, memorization and logical analysis; and especially their powers of intuitive precognition and elaboration. These adaptations did, in fact, occur and, consequently, new cultures and societies emerged, and so did the concept of time and space, as humans remained in place long enough to notice the changes in the local natural environment over repeating periods of time... time itself could now be segmented into *phases and cycles*. And, the same skies could be viewed from horizon to horizon and now presented recurring movements of the Sun and the Moon, as well as Venus and all the other bright lights.

The *big picture* reality of the effects of external forces on local weather became ever more apparent as the attempts at settled agriculture proceeded. The farmers of the *New Lithosphere (Neolithic)* eventually became aware of the distant derivative effects of global warming on their local agricultural fields and their shallow-fishing waters.

The sedentary colonists also began to recognize repetitive and cyclic aspects to the patterns of rainfall; to the alternating periods of dry spells and wet seasons. They gradually came to appreciate the finer coincidences between botanic lifecycles and the changing seasons, and more particularly, in the congruence of Sun and waters at the most propitious planting and harvesting times.

And eventually, there came a time when the individual memories became collectivized in a *cultural memory*. And over long-count periods of observing the same skies, night after night, the skywatchers continued to memorialize these observations... at first through oral traditions and later through marks and abstract symbols – impressed on clay tablets or chiseled in stone.

As individual data points of experience became evident patterns, some Enlightened Minds began to detect "causalities" that affected the performance of their agricultural system. In the beginning millennia of the Holocene Period, these were thought to be a result of human behavior in some way or other, and therefore, could be controlled by adherence to certain algorithms in daily life.

But even with the strictest adherence to rituals and rules, there still would continue to be weather-related natural disasters that would introduce perturbations and even chaos in the food-production systems. These were mostly manifested in the form of disruptions in the expected seasonal flooding of the rivers and lakes... or some other major hydrological occurrence.

But especially during the early millennia of the Holocene period, there would still be some major and many minor tectonic events, ripples from the earlier planetary changes... earthquakes, tsunamis, floods and volcanic eruptions still would cause regional or local cataclysmic effects on settled agriculture. And, then there were the relatively rare, but greatly destructive impacts of asteroids, comets, meteoroids and other rubble left over from the construction period of the solar system.

Then there were the ripple and rebound effects of the shifting of weight on the surface of the planet, as the released waters of the melting ice caps flowed into the oceans. There would still be tectonic perturbances, such as earthquakes, volcanic eruptions, tsunamis and the landslides well into the Holocene Period. The most significant of these disruptive events were the flooding of the Black Sea by the invading waters of the Mediterranean Sea about 6,000 years ago, and the volcanic eruptions that destroyed the Minoan Civilization at about the same time. Together, they would ultimately be conflated and remembered as the *great floods* in many human myths and legends.

These continuing cataclysmic events under and on the surface of the planet signified the final perturbations on the lithosphere, which were caused by the shifting of the water-load from the polar ice caps to the oceans, as the melting ice water sought its lowest level. The net effect of all this secondary remodeling was the finer redrawing of the coastlines and riverbeds, which represented the most sought-after real estate for the Holocene colonists who needed accessible and constant or at least cyclically reliable water sources for irrigating their fields, for making sun-dried or oven-baked bricks, and for the personal needs of the burgeoning populations.

So, by about 8,000 years BCE, the initial agricultural colonies sprung up along the Yellow River (China), the Indus River (SW Asia); the Tigris-Euphrates Rivers; and the Nile Rivers of Sudan and Egypt. Other agriculturally-based colonies were established along the coastlines of the major inland seas; most prominently the Mediterranean and Black Seas.

Chapter 13

New Beginnings

The Pleistocene *time of troubles* had scattered some human population centers and isolated others. The flows of refugees fleeing various catastrophic events took a myriad of directions throughout the Tropic of Cancer Region, over many millennia and generations. By 30,000 years before the present time, however, the descendants of the survivors had finally established new human regions... once again, near the waters. It was the beginning of the second planetary colonization of Earth.

This new period of human history began about 10,000 years ago; a new human region was developing on Earth. it encompassed the riverine valleys of Mesopotamia and the valley of the Nile River; the lands around the Black Sea and the Mediterranean Sea; the major rivers of China and India; and the Yucatan Peninsula of America. These would become the major nodes of a natural and cultural region of *milk and honey*, where the sun-soil-water variables were especially propitious for settled farming. It was in these places that the new Holocene colonists would prepare for the next wave of migration and colonization of the planet Earth... and for the ultimate colonization of the Solar System.

In each of these discrete places on Earth, a new model of sedentary agriculture appeared about 10,000 years ago, as the Holocene humans began to recolonize some areas within old Tropic of Cancer Region; and created new colonies in the higher latitudes of the planet, and in the Americas.

What the human mind had first imagined was now a reality; a new construct of a food-producing emerged. It transformed *hunting grounds* into plots of land whose boundaries and content were designed by humans. These were *fields* which contained determinative mixtures of living soil and regulated waters.

They were essentially food factories, which worked intensively by their human caretakers, to produce a desired packet of calories and nutrition. In most cases, the predominant plant was a progeny of wild grasses, from which a cereal would be *domesticated* to produce the optimum caloric yields. Ultimately, this would mean having to develop the right combination of purposed *fertilized* soils, Sun, and water on a larger ecological scale.

Enlightened Minds

It was in these natural laboratories that a new generation of Enlightened Minds received the insight to understand how the decay of plants and animals, and the excrement of animals was part of a symbiotic relationship whereby the living plants were reinvigorated. Ultimately, this realization led to an intimate *familiarization* with the life cycle of plants and animals to produce food and fiber… There was also a realization of the other variables that made these fixed-units of land viable as food-production *factories*, such as variable solar radiation and controlled inputs of water.

WITH THE INCREASING POWERS OF OBSERVATION, THE ELUCIDATION OF PATTERNS, AND THE APPLICATION OF LOGIC... THE HOLOCENE HUMANS CREATED A MORE POWERFUL AND SOPHISTICATED SYSTEM OF KNOWLEDGE, SKILLS AND TOOLS... THESE WOULD BECOME THE SCIENCE, ENGINEERING AND TECHNOLOGY OF THE SPACE AGE...

Early Holocenes utilized their advancing suite of K-S-T (Knowledge-Skills-Tools) to create a new model of intensive and systematic agriculture. Succeeding generations of their progeny would apply this basic paradigm to create new models of manufacturing, transportation, communication… etcetera, etcetera…

The diversification of economic activities also generated basic changes in the culture and societies within these central places. Within these spaces designed by humans, there were those dedicated to certain specialties, such as the manufacture of tools and implements, which were used in cultivation and in husbandry. These specialized spaces had existed to some extent in the central places of the Pleistocene period, but these were much grander and became more sophisticated as the Holocene revolution continued.

Some of the societies that developed along the rivers and around the lakes proved to be more successful than others – usually due to a *comparative advantage* – which derived from being located on more fertile soils or because of their relative location to the more reliable water resources. Others drew their comparative advantage from their serendipitous access to primary materials, such as a certain kind of stone (after all, these were still *lithic* places) or an essential element such as salt, or a desirable mineral such as gold, silver, copper, tin or iron. And then there were the societies whose comparative advantage consisted of having a certain *critical mass* of Enlightened Minds in their populations.

Eventually, the more successful places attracted many folks who did not engage in farming or mining or any other *primary activity*. Nor did they *manufacture* anything from the product of the primary activities. These were folks that worked in the so-called *tertiary sector* of the economy; they produced *information* and *ideas* which were the catalysts for the development of the earliest human civilizations. These emerged with the appearance of a new class of "idle thinkers" who drew inspiration from the Cosmic Mind… which was channeled through Enlightened Minds.

These Enlightened Minds began to appear as early as 10,000 years ago, in certain riverine and lake civilizations: in Mesopotamia and in Egypt; within the Indus River system in Pakistan and the Yellow River region of China… as well as the *lake* civilizations that emerged along the coasts and on the islands of the East Mediterranean Sea and the Black Sea.

About 5,000 years later, the northern frontier territory of the Fertile Crescent Region began to produce more advanced civilizations… within the domains of the Aegean Sea and Asia Minor. It was in these new human colonies that unusual clusters of Enlightened Minds began to appear, especially after about 1,000 BC. It was in this region that many of the Enlightened Minds of the time developed much of the intuitive, idealized product (philosophy) that would ultimately inspire later Enlightened Minds of the periods known as the Age of Enlightment and the Renaissance, which – in turn – inspire the Age of Positivism and the Space Age. Indeed, it can be said that the preparation for the colonization of the rest of the Solar System began in central places like Ionia (Greece) and Alexandria (Egypt), which were iconic central places of the Hellenic Civilization.

Operating as the "markets" of information and analytical techniques, these central places produced the archetype "philosopher-scientist" of the early Holocene period. It was in these centers of human minds that the accumulated knowledge of all previous Enlightened Minds began to be collated and organized into models of the real world, which were proposed to describe and explain all manner of physical reality.

This is where the *a priori* knowledge of the atom; of matter and energy; of gravity and electromagnetism; and of the planets of the solar system and of the stars that illuminated the skies began to be rationalized and expressed.

It is where the process of learning itself was subjected to careful analysis and produced the competing schools of thought known as empiricism and rationalism. And, it was from these central places of the human mind and brain that growing Cosmic Library of Human Knowledge – would illuminate the Enlightened Minds of the age of space exploration and space colonization.

The Enlightened Humans were individuals who dreamed and mused about various aspects of the physical world, some of which they could only intuit with the mind, but also many others that they would transmit to the brain for *scientific* processing. In the end, the combination of intuition and disciplined conscious analysis would produce a tradition of *structured imagining* that has furthered the "long-count" objective of colonizing the solar system.

Meanwhile, modern humans have continued to develop and improve tools for detecting and observing phenomena that cannot be *seen or felt* by the normal senses of the brain-body system. An example of these are the telescope and the microscope – both of which have served to enhance the innate powers of the eye-brain system. Still other devices have been created to enable humans to observe wavelengths and spectra of light and sound that cannot be *seen or heard* by the unaided human eyes and ears.

The same can be said for the innate *eye* of the human mind… whose powers have sometimes been enhanced by drugs and meditation. Stated succinctly: throughout the Holocene period, the human mind has been developing greater powers of imagination and projection in time and space that extend beyond the actual physical realties. This higher level of cognition and reasoning has been manifested by increasingly abstract communication from the mind, which is manifested in the spoken and written language, as well as the more precise, yet abstract language of mathematics, and artistic expressions of the human mind. It is through the mind that the intuitive spoken and written, as well as artistic, communication continued to be developed… ultimately to appear as the fundamental language of the computer, whose "alpha-numeric syntax would be expressed as a series of "photon packages" of zeros and ones.

One of the most significant outcomes of the writing and record-keeping is the emergence of patterns; and the realization of similarities that is often the embryo of a new way of looking at phenomena of the real world. Thus, it was the repetition of symbols that describe phenomena, or to keep track of the output that summarize a series of experiences, which then begin to leave patterns and matrices in the memory of individuals and in the social groups of individuals.

So, an important power of the human mind is its ability to recognize patterns, along both the dimension of space and time. These patterns or mental maps, which exist only in the mind, are often used as blueprints for creating human regions on the surface of a planet or in interplanetary space. With these new abilities to construct a human region solely in the mind... and to employ them as a blueprint to create a human spatial systems or regions, where nodes and linkages.

In the Early Holocene period, the nodes were homes and farms; their factories and shops; their stables and compounds of domesticated animals... all interacting in pursuit of an intelligent objective.

Applying "Know-How" and "Can Do"

In theory, an adequate base and lever, can multiply the force that is being applied to moving or lifting virtually any massive body. An analog of this observation might be: "give me a storehouse of common knowledge, tools and skills to draw from, and with the lever of logic, I can learn about all phenomena in the universe." More generally, one can say that analog thinking is a lever which multiplies the power of induction and deduction in solving new problems. This "cultural leverage" has indeed served to enhance the power of learning and adapting which has enabled both Pleistocene and Holocene humans to make the continuous advances that have been realized in the colonization of planet Earth.

By 2000 BC, the human brain had by now become fully developed as an organ of *sentience, awareness, logic, control and memory...* in relation to its internal systems and the external environmental systems. And like the artificial intelligence devices of the 21st century, the human brain became exceedingly good at performing the tasks related to sensory data capture, the validation and processing of the captured data, and the projection of information and commands to the actualizing the actuator subsystems. The development of the human brain systems has been essential to human progress, but the most powerful tool for colonization of the Solar System is the human mind. This is the interface between the Cosmic Mind and the human minds, through which the Holocene Homo spaciens receive the guidance that is needed to carry out the colonization or new worlds.

Since the second Great Human Awakening, which began about 7,000 years ago, most of the illuminations have been received by Enlightened Minds in the "Old World". This was the era during which the natural environment and the heavens above became the subject of careful observation and logical reasoning. It was the first time in human history that the human brain was fully-equipped to process the growing stream of data it was receiving about the physical world from its increasingly powerful and sophisticated external sensors. It now also had the necessary power of perception and cognition to evaluate data before sending it to its logic centers to produce useful information. It now had the memory to place information into proper context… for solving problems related to its own survival and that of the species.

It was during this period of human development that the secrets of planet Earth began to be probed. Fire and water; gravity and electromagnetism; air and aether; and the celestial bodies were subjected to intuitive and empirical analysis. Also, the language of the universe, mathematics, began to develop as a logical key to understanding the Cosmic Context.

And, beginning in the 18th century CE, these flashes of illumination are visited on humans who reside in the Americas and Australasia too. In all these places, modern Enlightened Humans began to concentrate their attention to the empirical study and rational understanding of our portion of the Cosmos.

This marked the beginning of space colonization within the human mind…

Geospace 2060

OVER THE MANY THOUSANDS OF GENERATIONS OF HUMAN DEVELOPMENT DURING THE HOLOCENE PERIOD, THE INNATE AND ENHANCED POWERS OF THE MIND HAVE BEEN USED TO CREATE A POWERFUL AND ABSTRACT MODEL FOR COLONIZING THE ENTIRE SOLAR SYSTEM...

In a static form, this abstract model is presented as a construct of nodes and linkages. In its dynamic form, the nodes are centers of constant activity and interaction. The linkages represent flows of people, goods and information. These flows bilateral and multilateral. And their intensity represents the cultural energy of the human spatial systems.

The construct of nodes and linkages create an interactive and interdependent spatial system, or region. The nodes can represent individuals or groups of humans, both of which can represent nodes of a spatial system, at different scales. The nodes are then interconnected by linkages to form a network. In this scheme, there are certain nodes which experience the highest traffic and develop the highest concentration of cultural energy.

Chapter 14

The Calculus of Space Colonization

Throughout the Solar System, the availability of water is the essential factor in making the decision of where to locate an earthling colony. Water, aside from its internal utility, is also a source of oxygen and hydrogen, both of which can be used to provide propulsion energy or to provide power and light for colonies. Water is also an essential agent in many natural and artificial processes that occur throughout the Geospace Region.

Water is universally reliable in its properties and behavior. This means that all water always behaves within the parameters of the laws of physics, varying only as to the effects of pressure and heat or cold… or of gravity, electromagnetism… or the subatomic dynamics of natural matter and energy that occurs everywhere – whether on celestial bodies or aboard artificial spacecraft. Therefore, human colonists can expect that Holocene knowledge and skills regarding water can be efficaciously applied throughout the rest of the solar system.

The availability of accessible energy resources is another important factor in deciding where to establish an earthling colony. Fortunately, like water, energy is ubiquitous throughout the Solar System, in one form or another. Like water, energy occurs in many states of being and it too behaves according to the laws of physics.

Energy is light, and light moves through space at nearly the speed of light. It moves in the form of waves, of various lengths, and at various frequencies as it passes a given point of observation. Energy waves also vary as to initial power and distance-potential. These properties have proven to be useful to astronauts and colonists in space, including as a means of observation and as a medium for transporting raw data and finished information.

Energy propagates, flows and radiates… in the form of light and heat, or as plasma and other entities of charged particles. And transmitted energy often collides with magnetic fields and plasmas of charged particles. The energy that is transmitted from the Sun or the stars and from radioactive decay also propagates throughout the Solar System, and interacts magnetic fields or of charged particles of planets, moons and other celestial bodies.

Most important for space colonization, energy behaves the same everywhere, and it has the same effects throughout the Solar System. That means that human space colonists can expect to treat and utilize energy effectively with the familiar suite of knowledge, tools and skills that have been developed throughout the Holocene period on the home planet.

So, space colonists can expect energy radiations to have negative "weathering" effects on all natural and artificial entities, and on the earthlings and their machines and electronic systems. On celestial bodies with robust outer fields of charged particles and significant atmospheres, the cosmic, solar and nuclear sources of radiation are mitigated or otherwise affected by a variety of natural shielding mechanisms. These layered envelopes of protection, including the atmosphere, the magnetosphere, and various other fields of plasma protect "living" celestial bodies from the weathering and degrading effects of various forms of charged particles and the blasting effects of dust particles.

No light, not even sunlight, is completely benign. Thus, prolonged exposure to cosmic and solar radiation can be dangerous to humans, especially their outer organs, the skin and the eyes. For example, acute exposure to the Sun can cause skin cancer, cataracts and blindness, unless some form of artificial protective shielding technologies (e.g. sunscreen, clothing, shelter, or sunglasses) is used.

On Earth, living beings are shielded from cosmic and solar radiation by the combined properties of the magnetosphere and the atmosphere. But In outer space, beyond the geomagnetosphere, humans have discovered a vital need for artificial shielding of cells, molecules and atoms from the negative effects of all sorts of *precipitation*. In outer space, "precipitation" refers to flows of electromagnetic and atomic radiations.

All these forms of *precipitation* have required human astronauts and colonists to develop *shielding* technologies and techniques to prevent damage to biological cells and molecules of materials and electro-chemical processes in space. Radiation emitted by stars and the nucleus of atoms affect both organic and non-organic structures and processes: it degrades or changes organic and non-organic systems; it causes "weathering" of materials in much the same way that variations in temperature combine with water and wind to relentlessly destruct mountains and buildings on Earth.

Thus, radiation of all types become more significant to humans as soon as they leave the protection of the combined natural shield of Earth's atmosphere and magnetosphere. Once in outer space, humans and their artifacts face the relatively unmitigated effects of cosmic and solar radiation, as well as the radiations resulting from the aftereffects of cosmic construction and the continuing decay of the elements.

Some other planets are large enough and still active enough to generate their own magnetospheres, either because they have a core dynamo or have magnetic fields that are embedded within their lithosphere. But human colonists on celestial bodies with no natural shielding, and human colonists aboard orbiting space colonies must develop artificial shielding systems that can approximate the protection of natural magnetic or charged fields. These can be either "passive" or "active" systems – or a combination of both.

Passive shields involve the use of natural (soil and rock) or artificial materials (e.g., metal alloys and plastics). Fibers, both natural and artificial, can serve as efficacious shields for both individual humans or their infrastructures and systems. On many rocky celestial bodies, such as Mars, the Moon and the larger asteroids, the primary shielding strategy is to go underground; to use the natural shielding properties of rocks and soil.

Active shields usually mimic the operation of natural fields of charged particles, such as planetary magnetospheres and plasma fields. Such *force fields,* consisting of charged particles, are used to counter the incoming flows and clouds of electromagnetic and nuclear emissions from the stars and our own star, the Sun, as well as from the structural changes within atoms and the related emissions of neutrons and charged electrons.

Chapter 15

Proactive Colonization

At the beginning of the Holocene Period, a new generation of Earth colonists had created a new paradigm for surviving and thriving on a changed planet; that is, to "following the water" as their basic modus operandi for colonization. This time, though, they would develop a new system for actively engineering the water systems to "manufacture" their food and fiber, by intensively cultivating a specific plot of land, and artificially intervening in the natural growing cycle of plants.

It was also during this time in human history that the efficacy of linkages increased in tandem with the burgeoning nodes in the developing human region. Thus, there was an engineering of these linkages in the form of roads and canals, as well as the virtual neural pathways of knowledge, which created the spatial human regions. And, along these linkages there were migrations and cross-migrations, as groups of humans effectively recombined the genetic material that had been fragmented and isolated during the end-times of the of Pleistocene Period.

Holocenes also learned that successful human colonization, everywhere requires adequate supplies of water and food, and a certain balance between internal and external pressure on the human organism. They also learned that the paramount factor in determining the success of a human colony is its human resources, and that the capabilities of the human brain-body system and its developing auxiliary *artificial intelligence* entity – the human mind are dependent on these resources. These are what give *life* to the natural resources and which determines which of the colonies will be more *successful* than others.

Looking back, although the nodes of the Pleistocene Period had been continuously mobile, they were nevertheless connected by dynamic linkages. The clans were continuously on the move, but they occasionally paused for a limited time at certain places, where an exceptional abundance of water, game and plant life made them good *rendezvous* sites – where the various clans of a region could refit and resupply; and interact and exchange genetic material along with ideas and technologies… to advance the common quest for planetary colonization. The colonization of the planet appeared to be moving along a successful trajectory.

But, after about a million years of colonization of planet Earth, the Pleistocenes found that the planet and its natural environments, as they had known it over many thousands of generations, was steadily and imperceptibly trending towards a period of cataclysmic changes; the very ground on which the various colonists learned how to survive and thrive was undergoing a period of extensive renovations. The existing "long-count" Ice Age was ending, and the dependable times of the water availability and expected temperature regimes were being disrupted, and ultimately began to significantly disrupt the Pleistocene societies – some more than others – depending on where they lived on the planet. However, even as the Late-Pleistocene colonists were being confronted with existential changes of their natural environments… they persisted in the traditional wandering, hunting-gathering paradigm… and that would ultimately spell the end of their time as a viable species on Earth.

Geospace 2060

Déjà Vu`

Many thousands of generations later – in the age of machines and fossil-fuels – another generation of Holocene colonists would again experience a fundamental climate change on planet Earth; once again there were planetary changes that threatened the existence of the human species. Thus, global warming would provoke tectonic disruptions of the mantle and surface of the planet; this time, however, this new generation of humans would have the power to worsen these natural effects by their sheer numbers and with their greater powers of science and technology. Fortunately, this time humans would have the ability to escape the troubled home planet, and to migrate and settle on other parts of the Solar System.

However, they also faced the reality that any human colony that is established outside the planet of Homo sapiens must either import the basic elements of life from Earth, or else find ways and means to formulate them or to extract them from the natural chemical elements available to them in outside earth. In outer space, this can only be done through significant applications of knowledge, skills and tools.

And, space colonists are now applying the lessons they learned at home during the last two million years, as they endeavor to carry out the colonization imperative. Thus, the choice of location of any human colony within the interplanetary medium or on the surface of celestial bodies is primarily based on the availability and accessibility of existential resources, such as oxygen, water and food, through the deployment of artificial technologies. Beyond that, they understand that the efficacy of human colonization within the interplanetary space and on celestial bodies requires analog *atmospheric pressure regimes* and *earthly gravitational forces* within the space colonies.

Therefore, for successful "permanent" existence in the extraterrestrial environments – the ways and means; the technologies that include the *domestication* of the chemical elements, earthly processes, nanotechnologies, micro processing, 3D printing, and so forth... must be integrated as standard systems *onboard* the space colonies... the cell, atom and photon must become the ultimate renewable natural resource; with these capabilities, each colony can be an autonomous node in the spatial and functional system. Ultimately, every space colony must be self-sufficient, while simultaneously being capable of interconnecting with every other space colony.

And so, as had been the case with Holocene colonies on Earth, the space colonies have also become more diversified in many ways. Some space colonies are luckier with respect to relative location to asteroids and comets, which can provide them with water and other natural resources – with less costs because of less gravitational friction in transportation. That is, they have comparatively easier access to natural resources than other colonies within larger constellations of colonies.

Again, it is this variability and the inequality of comparative advantage that produces the impetus for trade among the space colonies. It is this impetus to trade products and services that is the driving force for developing inter-colonial political and social networks throughout the emerging Geospace Region.

This can be seen in the development of certain space colonies whose primary function is to extract primary materials, such as water ice and minerals from "nearby" planets, moons, asteroid or comets. Other space colonies are equipped to process the raw materials for distribution to the space colonies which manufacture finished products for the entire space economy.

Nodes and Linkages

The elaboration of the nodes in spatial systems inevitably cries out for the development of an equally elaborate system of linkages to connect the nodes, and thereby form a functional region; to create a system of interacting and interdependent parts, working together to achieve a defined objective or purpose.

Linkages always connect nodes within a spatial system, and sometimes connect to *exo-nodes* that lie beyond a spatial system. Linkages can facilitate or impede movement, depending on the variable properties of the Linkage System (LS) or of the Carrier-Payload-Propellant System (CPPS).

The LS can occur on over land or across bodies of water, and most recently, within the atmosphere, through the atmosphere and through the interplanetary medium, and now, into the interstellar realm. Regardless of the dimension of space that is being traversed, the crucial factor that applies to all linkages is that of "effective friction". This can refer to the basic effects of gravity or to the capabilities of application of external force to effect acceleration of a body. The efficiency of a linkages also is affected by the net mass of the carrier, its payload and the propellant.

Chapter 16

Earthlings in Space

Humans are *earthlings* in every sense of the term. They are of the planet Earth, and they are designed by nature and nurture to live within its environmental parameters. Thus, they require oxygen under very precise conditions; they must have water to hydrate their internal systems and to properly regulate body temperatures; they require a minimum quantity and quality of caloric units to power the brain and the body; and the body needs just the right balance between the internal and external pressure.

Outside the geomagnetosphere, hardly any of these life-support imperatives exist naturally. To survive under the literally inhuman conditions in the rest of the Solar System, the elements of the *Goldilocks Zone* must be imported from Earth or else fabricated with human technologies onboard the space colonies.

Consider that the anaerobic conditions of interplanetary space are approximated at certain places on Earth. Thus, at elevations of 12,000 feet above sea level, or at altitudes above 12,000 feet in the atmosphere, or under the surface of the waters, earthlings suffer from insufficient oxygen to maintain life without some technological augmentation. However, on some celestial bodies, most notably on Mars, the differences in environmental conditions, from Earth is often one of gradation, rather than absolute difference. Thus, there are places on Mars where temperatures are sometimes not excessively cold when compared to the Antarctic; and not much dryer than the Atacama Desert of South America.

Therefore, one way in which earthlings have been preparing for the colonization of the alien regions of the solar system is to seek out natural environments on other celestial bodies that are somewhat analogous to natural environments on Earth, such as on Saturn's moon, Titan. These are many places in the outer Solar System where the natural environment at least approximates some places on Earth: where there are deserts and the mountains are alike; where the atmospheres are somewhat similar to the stratosphere and the ocean trenches or the tundra; where the polar regions of the terrestrial planets and some of the moons of the Gas Giant planets contain methane or ammonia swamps that resemble conditions in Earth's water swamps. The upshot of the matter is that these environmental types throughout the Solar System can be used as laboratories to train earthlings and otherwise prepare them for the *inhuman* conditions in space, either by genetic mutations or epigenetic technologies and skills.

Humans on Earth perceive gravitational forces from a geocentric and familiar perspective, most commonly when walking, manhandling heavy objects, or hurling rocks or projectiles. Beyond that, special effects from so-called 'g' forces are felt when external forces cause sudden and acute movements of the head and body.

By the 19th century CE, earthlings were experiencing other dimensions of gravitational forces, as they began flying through the atmosphere, and as they began challenging the power of gravitational forces to achieve *escape velocity*... into outer space, where gravity takes on a solar-centric reality.

In space, gravitational forces "feel" alien to earthlings, either inside spacecraft or during *extravehicular operations*. And on celestial bodies, the interaction of mass and the distance from the Sun or Jupiter... or the combined mass of nearby celestial bodies, combine to generate other permutations of earthly gravity. Nevertheless, the gravity wells of massive celestial bodies present special gravitational challenges that must be overcome if earthlings are to be successful in colonizing the other planets and celestial bodies of the Solar System.

Ever since Homo erectus descended from their arboreal paradise – following the Great Human Awakening Event – they and their descendants have encountered events that threatened the continued survival of the species – first on Earth and later from cosmic and solar events in outer space. The most effective way that humans have been able to respond effectively to these existential challenges has been to adapt, both organically and culturally; that is, by using technology and social organization.

Consider that had the earliest Homo erectus and their cousins retained their primitive ape-like brain-body system, these earliest humans probably would have been doomed to an early extinction as a species on the savannas of Africa during the incubation period. Fortunately, however, along with the command to propagate the heliosphere with human colonies, there came a set of programmed instructions from the Cosmic Mind to guide them in the ways and means to exceed the innate capabilities of the earliest *naked humans*…

Fortunately, as it happened, the early Pleistocene humans – the first colonizers of the home planet – literally "hit the ground running" following the GHAE. So, in a matter of a relatively few generations of genetic and epigenetic adaptations, they had effectively reconfigured the erstwhile ape-like, tree-roving body, by restructuring their skeletal system and their sensors, to fashion an optimum *planetary rover system*. Equally, importantly, they developed a more powerful and efficient brain; one which could develop as an optimum *control center*… for responding efficaciously to any changes in the environmental matrix… that threaten human survival and propagation. One that could literally see beyond the trees…

The Mind Develops

The Holocenes later concentrated on the development of an entity to develop greater precognition and powers of imagination and intuitive logic, even as they continued to advance the powers of analysis of empirically-acquired data points from the external environment. So they proceeded develop the powers of the human mind – to create visionary solutions to the challenges of planetary colonization, mainly by developing the intuitive and *a priori* powers of the human mind; to create a mental imagery of the Solar System, even before the brain gained awareness of it.

The Holocenes also developed algorithms and information systems for terraforming place within the interplanetary medium and on the extraterrestrial celestial bodies of the heliosphere. One of the ways they did this was by organizing and mapping the data and imagery derived from both sides of the brain. These are then submitted for analysis by the brain-mind system; and by a rigorous testing of the product of the brain-mind processing, until a significant degree of successful outcomes is achieved in the real world. They also did it by developing a general suite of algorithms, which could be utilized in the establishment of a colony anywhere within the Solar System, including how to: (1) select a "suitable site" for the proposed colony; (2) determine the optimum absolute and relative location of the colony; (3) determine its appropriate scale, architecture and configuration; (4) develop the internal systems, especially those that can create a replicate of earthlike oxygen, pressure, gravity and temperatures regimes; (5) developing the optimum water and food systems; and (6) develop adequate power, light, and other utilities...

Aliens on Earth

The human sense of alienation from natural aspects of the planet Earth began during the middle of the Pleistocene Period, and continued to grow during the Holocene Period. By the time the first constellation of orbiting space colonies had been established, the human engineering of the cell and the atom had developed to the extent that the ratio of births and deaths had resolved the Malthusian paradox on the space colonies. This was the most definitive manifestation of human feeling of apartness from the natural world.

As if emphasizing this alienation with Earth, humans began to turn their eyes upward. They began to persistently and persistently study the skies; to take note of the movement of not only the overhead clouds, but also the distant heavenly bodies... and to then connect that "distant" phenomena to mundane phenomena... such as patterns of precipitation and temperatures and the other variables which affect agriculture and other aspects of the human society.

One of the first phenomena they noticed in the skies were the distant stars and the nearer Sun, the Moon and the planet Venus… which appeared variously in the morning and night… from one end to the other of sky watcher's field of view… as they appeared over one horizon, transited the skies, and descended beyond the opposite horizon… leaving the world in darkness. But even within the field of darkness in space, there were points of light. They noted that the Moon and sometimes, Venus would appear as the Sun left the sky, and that they would oppose the daily movements of the Sun. The Moon also shone some light on the surface of the planet; softer than the Sun's and only at various times. Other points of light also appeared in what appeared to be the ceiling of the dark skies; they seemed to move about the ceiling in a kind of performance, which cyclically moved its venue in the skies…

These phenomena of the days and the nights seemed to appear and move about with great regularity; as if they were performing complicated dance routines. It was an awesome and mysterious display of the heavens to the earliest skywatchers; but eventually some Enlightened Minds were inspired to take note of and memorialize and movements of these celestial entities, and to seek out patterns… patterns which would be memorialized on fresh clay and as ciphers and abstract pictures drawn on papyrus… and as the sky watchers continued to keeping track of these cycles of phenomena, they would eventually crack the cosmic algorithms of the celestial heavens, and they began to create a chart of cycles past and, even future… In doing so, they added another dimension to the calculus to the colonization of the planets.

The ancient skywatchers could study the "ceiling" of the Celestial Sphere with only the natural powers of the human eye, the intuition of the mind, and the processing powers of the brain. They lacked even the most rudimentary artificial enhancing telescope as they studied the skies. But these ancient skywatchers turned this liability into a strength: they concentrated on developing their powers of concentration and obsessive persistence; the discipline to follow a standard algorithm for detecting new bright objects and for inserting them into a developing construct of their portion of the cosmos.

The extreme concentration of the skywatchers, over several hours, also served as a form of meditation which often transformed the observer's state of consciousness. It was in this altered state that the ancient Holocene skywatchers received flashes of intuition that would give them insights into the cosmos that transcended their innate powers of visual acuity. And, as the generations passed, these skywatchers took on the aura of a priesthood; as seers of the future and fonts of wisdom with respect to agricultural pursuits and military adventures too. Eventually, the priests of the skies would often join with chieftains and kings in governing the Holocene societies that developed along the rivers and around the lakes, and on the shores of larger seas and oceans.

As so often happens, these priests would require a special place for carrying out their activities. These included Stonehenge and at other places where special arrangements of large stones were aligned to capture the rays of the Sun on the solstice or equinox, or to isolate the brightness of Venus, or a given star. In the land of the Mayas in Central America and the mountain realm of the Incas of South America… massive stone formations and specialized observatory buildings were constructed, in devotion to the study of skies.

Fast forward to the 19th century; the modest telescope, which had been invented 300 years earlier, had grown so big that it required special structures and supporting infrastructures to do the work of watching the skies. These would be the iconic dome-shaped buildings with the giant telescope protruding from it like a cannon on a swiveling platform. And, relatively soon after, telescopes would be orbiting the Sun and planets to provide a form of *in situ* observation capabilities.

With these powerful optical instruments, the modern Holocene skywatchers – now called astronomers – carried out their studies of the ceiling of the Celestial Sphere. By now the ritual of careful observation had been converted to a new discipline called the scientific method. And the practices and rituals of the early skywatchers were subsumed under the art and science of astronomy.

Chapter 17

The Second Human Awakening Event

The second human awakening event began with the sudden awareness that the celestial realm, which heretofore had only been the subject of voyeuristic observation, was being observed and analyzed like any natural phenomena on Earth. And, by the middle of the 20th century CE, humans had become *familiar* with outer space, and now saw it as a new territory to be colonized. By then, astronomy had joined philosophy and theology in pursuing the secrets of the skies. The earlier schools of observation and thought had given humans an intuitive *familiarity* with the skies, but astronomy applied the powers of empirical observation and the scientific method to gain a more conscious understanding of the Cosmos.

The initial insight came when these skywatchers realized that what they observed on the ceiling of the celestial sphere were bodies that were like our own planet, in some way or other. Some Enlightened Minds intuited Earth was a moving platform, from which they were observing other moving celestial bodies, within a growing universe of moving celestial bodies. From this conclusion, came the realization that an observer on another celestial body would see the constellation of moving bodies from a different perspective. It reflected an understanding of a dynamic cosmos, which is constantly moving.

The earliest skywatchers saw the celestial sphere and the celestial bodies within it mostly with their mind's eye and less so with their eyes of the brain. Their comprehension was based more on intuitive imagery and less so on empirical data. However, the latter approach began to be used more and more by *philosopher-scientists* who observed phenomena and patterns, and applied logical analysis thought to gain a new familiarity with the celestial sphere.

Such "scientific" studies of the celestial sphere, based on empirical observation re-emerged in the 16th century CE with the development of the telescope. It wasn't yet astronomy in the modern sense, but it did represent a shift in trajectory towards it, and away from the theological approach, which had dominated the study of the celestial sphere for a millennium.

The first step was to identify and catalog as many celestial bodies as possible, within their restricted general archive knowledge and the visual acuity of the early telescopes. Nevertheless, within the existing parameters of information and technique, the early astronomers could construct maps and charts of the skies, from a variety of perspectives. Within this framework of knowledge and technology, what they still could not observe directly could still be intuited by applying analog logic.

As a result, the empirical data and the logical approximations of reality that has been achieved in the 20th century CE has enabled modern Holocene humans to gain a level of practical understanding of the celestial sphere; such that they can plan space missions with an increasing level of confidence; which is being heightened by the experience that is gained with each new space mission of exploration; and the development of new and more powerful tools of observation and analysis.

Even back in the 19th century CE, an interconnected global network of skywatchers was already developing, especially with the advent of the telegraph. Telescopic observatories were being deployed throughout the surface of Earth and, significantly, they were sharing data with each other. A later generation of skywatchers would begin the process of observation and data collection in support of space exploration in the 20th century. The next advance in familiarity with the celestial sphere also occurred in the 20th century with the expansion of the telescope's field of vision into other parts of the light spectrum invention. Astronomers were then able to "dial up" the best frequency of light to observe stars and other celestial bodies – in much the same way that radio operators "surf" radio frequencies to receive loud and clear sounds.

Chapter 18

The Space Imperative

During the 16th century of the Common Era, another constellation of Enlightened Minds emerged, throughout the Old World. Many continued to pursue the theories of the mind as dogmas; others, however, turned to the theories derived mainly (or ultimately) from the combined powers of the mind and brain. Followers of both paradigms, in one way or another, would be the creators of the science and technology that would support space colonization in the 21st century.

One important contribution of the followers of empirical science to the cause of space colonization is what might be called purposeful scientific modeling. That is, the construction of expectations of reality, which can be used as *the expected* reality in planning space missions. They are useful tools for identifying as many of the variables that are involved in a proposed space mission as possible; to develop an acceptable level of confidence in the prediction of reality during a space mission.

Other models seek to identify existing locations and movements; to use these data to construct probabilities of future locations and movements of one or many, if not all the celestial bodies (natural and artificial), based on careful empirical observations, comparisons with historical events and probabilities of future locations and movements of celestial bodies. It is really a matter of identifying and the entire population of cosmic actors, past and present, and to construct an elegant model that can predict the movements of all the celestial bodies – past, present and future.

An example is the geocentric model of the Solar System (Geospace Region) that is being used to plan all space colonization activities. It describes the movement of the planets and secondary celestial bodies in the practical terms that are needed to carry out space missions. It is also a quantitative model that utilizes mathematics to describe the structure and dynamics of the real world. It is a model that describes reality with the quantitative precision that is needed plan space missions, from launch to orbit, from orbit to landing, and at every intermediary point within the mission. And, within the "rules" of the model, the database related to the Solar System has developed; in the quantity of its content and in the linkages between data points, thus creating an ever growing "web" of knowledge and experience that could be used by succeeding generations space scientists. Meanwhile, surface-based observatories and those in orbit continue test hypotheses based on the Geospace Region model.

That humans had arrived at this point in the process of space colonization was in great part due to the generations of Enlightened Minds, who *lifted their eyes* from the surface of the Earth, peered through its atmosphere, and began to seek out the mysteries of the entire cosmos. They would discover the truths that would make it possible for later generations of Enlightened Minds to solve the mysteries of the physical world; of matter seen and unseen, and of the hidden forces of the universe.

Now, in the 21st century of the Current Era, the use of the telescope as a tool for skywatching has become ubiquitous and increasingly sophisticated. In part because of continuing advances in the fabrication of telescopes, and the benefits of mass production, "observatories" are increasing in number, and a greater number of eyes and brains are producing an explosion of data and a greater *familiarity* with the cosmos.

The telescope has now become a *skywatching system*. The tool itself has been continuously upgraded with ever more powerful magnifying lenses and reflecting mirrors. Other improvements to the original telescopes, which could only detect *light* within the wavelengths of *visible light,* have included: (1) electrical power to augment the natural brightness of light; (2) improvement in the effectiveness of the optical lenses and mirrors; (3) addition of sensors that can detect light throughout the entire spectrum of electromagnetic radiation; and (4) the connection of the telescope with the computer, which provides the capability for producing high-resolution imagery at the pixel level.

So, by the start of the campaign to establish human colonies in space, the venerable telescope had become a powerful, complex, and sophisticated system. Its assortment of sensors could now detect, observe, and track objects over time and distance to gradually peel back the layers of *dark matter and dark energy*.

Skywatching has become an applied science in the age of computers. It now has a purpose: to create *maps, catalogs, databases, and algorithms* of the Solar System in support of the objective of space colonization. Now the tools and skills of the skywatchers are used in the service of space exploration and colonization; astronomers now have become members of a constellation of new *astro-scientists and space engineers*. Together, these teams of Enlightened Minds are creating the knowledge and technologies to accomplish the cosmic imperative of space colonization.

But, with their new powers of observation into space, these same scientists also could peer deeply into the history of the cosmos, and they gained a greater appreciation for the fragility of the home planet.

> *"WE MUST CONTINUE TO GO INTO SPACE FOR HUMANITY..." WE WON'T SURVIVE ANOTHER 1,000 YEARS WITHOUT ESCAPING OUR FRAGILE PLANET...*
>
> --- STEPHEN HAWKING

The Second Time of Troubles

In the year 1945 CE, the first explosions of the Atomic Bomb occurred over two dense population centers in Japan. The immediate consequence of this man-made catastrophe was the local "extinction" of tens of thousands of humans – an event equal to that of the volcanic explosions such as those that occurred at Toba and Crete, both of which caused the extirpation of regional populations on Earth.

The more general consequence of these atomic explosions and the testing of the hydrogen bomb in the post-WW2 era has been the growing realization that unleashing the power of the atom and of hydrogen fusion could terminate the home planet as a viable place for human colonization. However, rather than reacting "rationally" in the face of such an existential threat to humanity, many nations have responded irrationally by making more of these weapons, to achieve a sense of security that was based on "mutually assured madness." At the same time, the proponents of the lingering Mutually Assured Destruction saga also saw the *need* to continue a new phase of the geopolitical struggle; a new battlefield in the low-orbit region of outer space. From this warped decision, humans finally began an upward into higher orbits, in the quest for the "high ground" above the planet Earth.

Historically, the power to cause the extinction of the human species on Earth has been ascribed only to a "Higher Power". And, for many, this belief would continue, even in the empirical face of the demonstrable evidence, which showed clearly that humans now had the power of extinction by the middle of the 20th century CE. And, even as many Enlightened Minds worked to develop solutions to the threat of self-emollition by humans, many others continued to ascribe the problem to their own activities.

So, another *time of troubles* on our home planet began during the 21st century CE. It was reminiscent of the Pleistocene end times, at least with respect to melting of the ice domains and the rising levels of the oceans. But this time, the natural events that constituted the existential threat were being abetted by human demographics and activities.

Thus, larger and denser human populations and the use of fossil fuels – and the actual use of thermonuclear weapons caused greater pain and instability and, at the same time, the options for dispersal and migration on Earth were more limited. Indeed, it was the recognition of these stark facts that finally impelled some Enlightened Minds to develop plans and systems for evacuating some humans into space; to establish new colonies outside Earth, as insurance for the continued survival of the human species.

So, as indicated earlier, the final impulse that initiated the space colonization effort was a *perfect storm* of multiple cataclysmic events on Earth. On the one hand, there was the was the global warming and its consequences. Then there was the catalyst to this warming caused by increased emissions of carbon dioxide from internal combustion and the burning fossil fuels. And, there was the thermonuclear fuse that could explode at any time. All these forces worked in unison to bring the human species closer to the brink of extirpation on the home planet.

Meanwhile, more "developing" human societies were becoming industrialized and they also increased the use of machines that were powered by fossil energy to produce the basket of manufactured *things*. To create the power to make machines and systems to do *work*, these neo-industrialists joined in the comprehensive search for and extraction of fossil fuels that were left over from eons ago. The upshot of all this was the overloading of the Earth's atmosphere with so-called *greenhouse gases* and the consequent trending towards a warmer planet.

Nature also added to global warming, as the planet again shifted its orbital pattern relative to the Sun and *wobbled* as it spun around its axis. Then there were earthquakes and other seismic perturbations; volcanic eruptions; solar storms… and other planetary maladies to the mix of global catastrophes that threatened the existence of the human species on the home planet. Thus, there was a continuous growth the human population on Earth, which increased demands for natural resources and added stresses on the planetary systems. A more ominous demographic factor was increasing growth and densities of human populations. The upshot is that every earthquake, volcanic eruption, tsunami… and nuclear explosion or core meltdown meant that greater numbers of humans would be killed or otherwise negatively affected by any cataclysmic event.

The likelihood of an apocalyptic event which could bring about the demise of the human species had therefore become an imminent reality. The underlying truth of the matter was that the home planet could no longer tolerate the burden of supporting the growing population of human colonists even in "normal" conditions. The demands on the global inventory of water and food had exceeded the natural capacities of the planet. Ironically, the *planet of the humans* was becoming almost as alien to the human colonists as any other place in the Solar System. And, always looming overhead were the asteroids, comets and meteorites, whose erratic orbiting behavior could cause them to collide with Earth and cause a greater magnitude of disaster for the human species. But, in the end, what brought the defining urgency to the matter was the proliferation of nuclear weapons, which was like a sword of Damocles hanging over the fate of humanity.

On the other side of the ledger, the course of events during the so-called Space Age eventually led to a viable path to a possible solution to the existential threat by adopting a strategy of dispersal and diversification as a way of ensuring the ultimate survival of humanity in space colonization. Dispersal in this instance refers to the migration of humans to other places in the solar system.

Within the overall objective of dispersal into other regions of the Solar System, the human decision-makers chose a strategy of space colonization which called for a multi-phase program of colonization. It also would be based on the notion of settling in places where the gravitational field is less profound; that is, where less energy is needed to launch, maneuver or land spacecraft.

In practice, this has resulted in the development of constellations of smaller colonies that have been placed in orbit around the Sun itself, or in orbit around the planets, their moons and the great asteroid Ceres. In the same vein, the Lagrange Zones that occur in several places within the interplanetary medium also have been selected as suitable sites for space colonies – in accordance with the "low mass" strategy of site selection. This strategy has enabled humans to soon establish a network, or system, of colonies that are dispersed throughout – rather than being concentrated only in Cislunar Subregion.

Also, it has been generally agreed that the marginal cost of establishing and maintaining human colonies in the outer Martian and Jovian Subregions are to be offset by the benefits derived from lowering the odds of a catastrophic event, which would threaten the overall survival of the human species in space. In the same vein of thought, having human colonies that lay beyond the Main Asteroid Belt and the Kuiper Belt (the home of comets) would increase the dispersal effect, thus lessening the probability of an apocalyptic strike by an asteroid or comet that would significantly impact the viability of the human species in space.

One crucial element that furthered the survival of the human species during the Pleistocene *end times* was the great diversity of the genomes, which had developed within the heartland of the Tropic of Cancer Region, but especially peninsulas and islands of Australasia, in which thousands of generations of humans had been marginalized and even isolated.

So, throughout the time of troubles associated with the end of the Pleistocene Ice Age, there was a series of migrations and cross-migrations which rearranged the human demography of the Tropic of Cancer Region. There also continued to be a series of out-migrations from the Africa homeland, northward towards Europe and Anatolia, and eastward into the Asian heartland and its peninsulas and archipelagos. But, simultaneously, there were also in-migrations from Eurasia, and back into a new heartland in the Fertile Crescent.

There was also considerable smaller-scale cross-migrations going on between Europe, the Middle East, Central Asia and North Africa… all of which added to the diversity of genetic material from which to carry out the organic and cultural adaptations to the melee being caused by gradual flooding and tsunamis; violent volcanic eruptions and slow encroachment of lava; and earthquakes.

Meanwhile, the diversification of genetic material in human groups was being accomplished by the growth of individual population groups; the greater the number of humans in a group implies a greater number of possible permutations and mutations in genetic material. At the same time, the mixing of genetic material throughout the Pleistocene and most of the Holocene periods of human history on Earth, occurred through the process of sexual intercourse. The propagation of the human genetic material has occurred as the number of humans in a basic unit grows too large and some members break off and form a new cell. Diversity in the gene pool has also been the result of the normal paradigm of the family, clans, tribes, nations and empires.

Genetic Diversity in Space

On space colonies, genetic diversity is being achieved through the application of genetic engineering at the level of the cell to create the optimum human organism for living and working in the various natural environment in space and on the celestial bodies of the Solar System.

One alien environmental challenge faced by humans occurs during longer space voyages and on orbiting space colonies. In all such situations, humans encounter cosmic and radioactive rays that are harmful and often lethal – sooner or later. This phenomenon of charged particles targets the human cell and the molecules of the materials that form the spacecraft and its various systems. One way of countering these hazards has been the use of genetic engineering technologies to heighten the resistance to the effects of radiation. Other technological responses seek to engineer organs of the body to function more efficaciously in anaerobic and microgravity environments. Still others superimpose artificial systems to enhance the natural capabilities of humans.

Another application of genetic engineering is aimed at alleviating the effects of physiological and psychological, as well as social stresses related to long periods of residence in outer space.

The Space Colonization Experience

Like the first planetary colonization in human history, which was carried out on Earth – first by Homo erectus and Archaic Homo sapiens during the Pleistocene Period and now the Holocene Homo sapiens – the colonization of the of the Solar System has continued; but not in a straight-line linear fashion, either over the surface of the planet, or over time. Rather, it has been a rather more or less coherent braid of linear movements that have been advancing amid the chaos that occurred towards the end of the Pleistocene and now during the 20th and 21st centuries CE... culminating in the current Space Colonization Program, which is a being carried out by a consortium of several nations and non-governmental organizations.

It all began with the development of Intercontinental Ballistic Missiles (ICBM) which were designed to transit great distances along the surface of the planet Earth. Their "orbit" either skimmed just above the surface (relatively speaking) or in hybrid orbits that began as vertical trajectories towards an "insertion point" in the stratosphere. At some point in the flight of the missile, the planetary gravity would bring it down to a targeted landing point on the surface of the Earth.

This type of intercontinental mission by the ICBM was intended to deliver a warhead on a target that existed on another side of the sphere. It would also become a model for planning missions in which the ICBM would become a space rocket, whose target would exist in outer space, and whose payload would be human astronauts, and reconnoitering and scientific equipment.

Central Places of the Mind

One reason for this consistent and purposeful drive towards the human colonization of the Solar System has been the appearance of "central places of the mind," in succeeding generations and at various points on Earth. These are the places where Enlightened Minds have congregated to develop the intuitive insights they have received from the Cosmic Mind, and to formulate proposals and plans for implementing them in the service of the imperative to colonize the entire Solar System.

Some of the earliest centers of the mind emerged as "schools" of philosophy (science) in the last millennia BCE. They appeared at a time when the when previous models of the world and the universe were being reevaluated, largely because of the ideas of the mind that had been presented by the philosophers of ancient Egypt. In the subsequent centuries, the Ionia Greek center of the mind would grow to encompass the broader Hellenic school in Alexandria.

The ideas of the Hellenic centers of the mind were maintained and nourished by the Islamic philosopher-scientists from 700 CE to 1500 CE. Thereafter, the centers of the Renaissance in Europe reemerged to add to the work of the earlier centers of the mind. They would be followed by the emergence of new centers of the mind, not just in the Old World, but also in the Americas.

The product of the work of all these centers of the mind would crystallize in the age of positivism, scientific method, industrialization, electrical communication, rocketry, and computation and control of artificial intelligence entities. And, ultimately, each of these centers of the human mind contributed to the coherence and stimulus of what would be the practical colonizing mission.

Historically, many of these central places of learning have been associated with universities or religious centers, but in the age of space exploration and colonization, they tend to be places where ideas have led to development of *ways and means, and of actualization… making things and making things happen.* Examples of the modern kinds of centers of the mind, which mainly arose in the last two centuries, are the national space agencies and the constellation of universities and private space agencies that interact to create a global web of centers of the mind.

These most recent centers of the brain and of the mind have no ivy walls, but they are often housed in sprawling campuses or within skyscrapers; many are not even concentrated in one building or one place; they are not always constructed of brick and mortar; and the *philosophers* and the pupils are located throughout the globe. Other, more specialized, centers of the human mind and the brain are found at the space observatories that have been established throughout the heliosphere. Their findings serve as cosmological datapoints for mapping the natural and cultural regions of the cosmos.

Telescopic observatories also probe the outer-space medium, the galaxies and the entire universe in search of points of light that that illuminate the current cosmos and those that were originally emitted as far back as the Big Bang event. The product of these data has also been propagated to centers of the mind and the brain for developing *information* that can be used by them in the quest to colonize everywhere.

The cumulative sum of all the knowledge and know-how acquired by all humans, especially by the centers of the mind, is now available to the modern space colonists. It is a *Cosmic Cloud* of human thought and wisdom; a powerful resource for planning and actualizing the human colonization of the Solar System. It contains digitized imagery and data, as well as a huge number of algorithms, for use in mission planning. These are presented in both analog (flowing) and digital (static) form.

One of the main contributions of the mind to space colonization is its power of envisioning a reality that exists only in the fourth dimension. It is a reality that is yet to be perceived by the human brain system, but which the human mind can perceive with its powers of *a priori* knowledge; it is a form of "artificial Intelligence," which was partly constructed by the neocortex; with instructions and blueprint provided by the Cosmic Mind. The upshot is that, from the advanced human mind, there has come the important concept of *Progress*.

So, the human mind has become the executive agent of *human progress* and, as such, the mind is the main locus of intuitive and creative thought and as the motivating force of human progress also means that all human advances in the colonization of the solar system are being directed by the mind. So now, the ideal of *progress* is inherent in the human spirit; it is what impels the advances in technology, science and social organization that serve to actualize the Cosmic Imperative. It is an integral belief in the notion that the human species is always moving towards new places and new situations. It is why humans began to migrate out of the birthplace of our species in Africa, and into the other continents and archipelagos of the planet earth for almost two million years – and it is what now impels them to migrate into the rest of the Solar System.

The cultural concept of *Progress* also impels the human species to continue developing as an organic and cultural entity, which can adapt to survive and thrive in every new natural environment. Onward and upward might be the most apt description of human development since the GHAE.

This concept of progressing along some linear curve has driven the quest to penetrate every type of natural environment, on Earth and throughout the Solar System. It gives coherence to each act of exploration… both in outer space and on other planets, moons and other celestial bodies, no matter how large or small, and even the most distant bodies of the Kuiper Belt at the end of the Heliosphere.

Geospace 2060

ALL HUMAN COLONIZATION OCCURS WITHIN SPACES AND SUBSETS OF SPACES... WITHIN SPACES THAT ARE RULED BY THE STRICTURES OF THE FIELD:

Chapter 19

Geometries of Space

Fields
Spaces can appear in one, two, three or even four dimensions...
A space is *field*... within which there are *laws* which regulate activities within it...

Grid Systems
Grid systems are logical devices for imposing order on abstract spaces; they offer a commonly-accepted language and logic for referring to unique locations

within a real space; they are logical constructs for referring to a specific point within two, three and four dimensional matrices…

A variety of functional grid systems can be utilized, but they all have a common purpose: to detect and communicate information about an object's relative location; its distance from any point of origin to any another point on the grid.

Points

A point can occur along a single line, or at the intersection of two lines which are not parallel to each other; or at the intersection of more than two lines that have different orientations.

> *[In space operations, a spacecraft can be depicted as a point, and its trajectory can be portrayed as a line].*

Points can move along trajectories, in any direction, within the dimensional limits of the space in which they occur. The movement of a point along a trajectory has the property of speed and direction.

Within a defined space, any point can become a reference point… from which a logical argument can be framed to determine static locations and to track the movements of objects within the space.

> *[In space operations, the position of a celestial or orbiting space colony can be selected as a central reference point, and from which lines can lead in any of 360 degree trajectories to any other point in space].*

Lines

Along a straight or curved line, the absolute location of any given point along that line can be described by a single reference (X). Multiple points can be described in the form of ($X_n X_2 X_n$).

[In space operations, there seldom occur straight lines; trajectories are almost always curved and orbits are a variance from the circle].

An intersection of two perpendicular lines can be described in the form of a coordinate (X, Y).

[In space operations, intersections usually occur as insertion into an orbit or as a rendezvous between two spacecraft].

An absolute point refers to the "resting place" of a point along a line (X_n).

[In space operations, there are no absolute positions; all points are constantly moving along a trajectory or within an orbit].

Moving points occur at various places along a line; at various times – as viewed by an external observer.

[In space operations, all points (celestial bodies and spacecraft) are always moving; relative to an external observer – at various times].

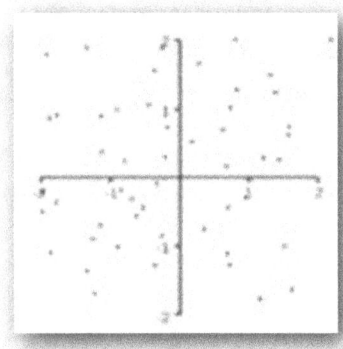

Two-Dimensional Colonization

Both the Pleistocene and the Early Holocene humans were effectively operating on a two-dimensional field as they carried out their global migrations of colonization; at least while they restricted their operations to the surface of the water and land areas of the planet. From any given point on these spaces, they could move along any number of lines… but only in a negative or positive direction, along a given line, and from a given starting point.

Throughout most of the history of human colonization on Earth, every space that has been colonized has been a two-dimensional plane, with only minor variations in height. So, in practice, colonists could only move along a linear path, in two dimensions. Therefore, it is understandable that they considered the planet to be a flat plane…

The first doubts about the validity of the flat-surface model of the planet began when Holocenes began to measure the movement of a given beam of sunlight along course of a daily cycle of solar incidence. The same measurement was done with respect to the size of the shadow cast by an object on the surface of the Earth.

Later, in the 16th century CE, some Enlightened Minds began to develop a new mental map of Earth – not as a flat plane, but as a sphere.

A New Mental Image of Earth

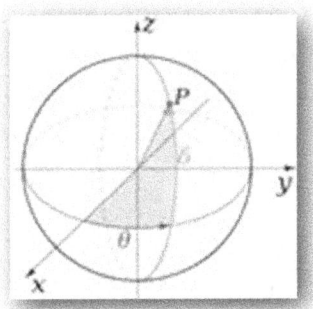

Once again, the human mind receives an intuitive flash of inspiration from the Cosmic Mind (a theory or model) and then passes the derivative hypotheses on to the human brain system for processing through the mechanism we now call the Scientific Method. That is: (a) empirical observation, (b) validation of data inputted by the senses, (c) development of a suitable logic construct for testing of the hypotheses, and (d) producing a validated model, or a more efficacious iteration of the model.

At the same time, Europeans began a new spurt of global exploration, which eventually led to a new layer of colonization in the Americas (this part of the planet had already been colonized by migrants from Asia about 30,000 years earlier). Some voyages of exploration that followed the rediscovery of the Americas would ultimately lead Europeans to circumnavigate the planet as they sailed farther and father to the south, and to the west or east. Thus, the spherical form of the planet was confirmed empirically as these missions of exploration continued to trace coasts of the Africa, the Americas, the eastern and southern littorals of Asia, and the archipelagos and islands of the Australasian seas.

With the recolonization of the New World by Holocene humans from Europe in the 16th century CE, came the practical acknowledgement of the spherical reality of Earth; it became a matter of common knowledge; a universally-accepted hypothesis that was proven by a series of voyages that effectively began and ended in the same place.

It was only in the 19th century CE that humans were finally capable of empirically viewing the evidence of Earth as a sphere, as they ascended into the higher atmosphere of the planet earth. First there was the balloon: a soft-material vessel filled with *lighter-than-air* gases that made the first extended excursion into the third dimension. It represented a new round of negotiation with Earth's gravitational forces... made possible by a growing familiarization with gaseous materials...

Three-Dimensional Colonization
The use of the balloon vessel, filled with lighter-than-air gases, would be the first instance of human movement into the third dimension of Earth; it was a precursor to human migration into a three-dimensional medium; and it

marked the beginning of the development of the technology for moving a massive craft through a medium by applying an external force (such as heated gas, explosive chemicals, or the propeller).

Thus, humans now could move in any direction of the compass within an atmospheric or an interplanetary medium. They now could defy the downward pull of Earth's gravity, but only while there was fuel for the propulsion systems that provided angular motion to keep the aircraft aloft, and in a state of dynamic tension with respect to the downward pull of planetary gravity.

However, as the limits of technology were extended, there emerged limits of human physiology and psychology…

Space Navigation

Because outer space is so large, and distances between both natural and celestial places are so great, space travel is a sequence of events, which occur in a series of time-places. Therefore, a fourth dimension is now required to describe location at any given **_TIME-SPACE EVENT._**

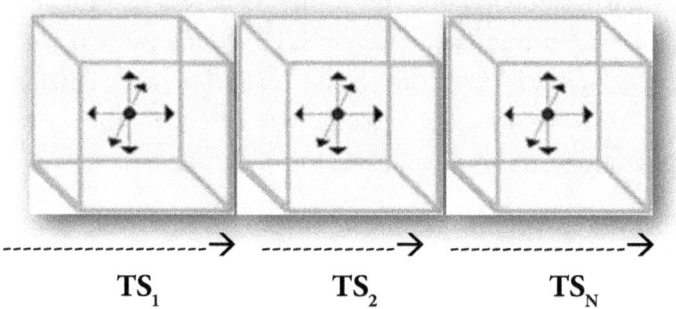

TS_1 TS_2 TS_N

Regardless of the number of natural dimensions that define a space, the human mind always projects an overlaying, intuitive *grid scheme* as an effective and efficient construct tool for describing the fixed location of an object, or to trace its movement along a line within any given time-space construct.

It should be noted that the two-dimensional construct can be used for both planes and spheres (the latter requires some *stretching* of the mind for longer distances). Within these logical boundaries, the two-dimensional construct is effective in dealing with surfaces of all celestial bodies throughout the Solar System.

However, within the *offshore* space which envelops a celestial body or within the interplanetary medium, a three-dimensional mental map and grid system is required for defining the absolute space-time location of an object. And, a four-dimensional dimensional construct is required to account for the movements of objects along trajectories, at high velocities, over longer periods of time.

So, there is no such thing as a fixed, absolute location for an object in outer space; all objects are constantly moving along an orbital path which is assigned to them by gravity; or one which is perturbed human propulsion systems; or by *natural externals*.

Geospace 2060

One mental map that has proven efficacious in portraying the constant iterations of location and movement of bodies around a central point within the heliosphere is the ecliptic; this is an imaginary plane which is projected in the mind's eye. It naturally extends outward from the Sun, but the human mind can select any other central reference point; such as Earth in a geocentric projection.

Colonizing The Third Dimension

During the latter decades of the 20th century, powered aircraft became the first "probes." They were machines whose mission was to explore the Earth's stratosphere... along a new frontier of the third dimension.

Later probes were sent into near-earth space to test the limits of humans and machines there; to collect *in situ* data about the frontier of outer space; to study the matter-charged particle-gas netherlands that help protect Earth from most of the ravages of normal solar radiation and cosmic radiation, as well as the steady rates of radiation produced by decaying or mutating atoms, and the occasional, acute storms of radiation produced by star bursts and solar storms...

And today, new generations of *astro-scientists* are sending robotic craft: satellites and orbiters, as well as long-count probes, to answer as many questions about the solar system as can be thought of by cosmologists, planetary scientists, physicists, geologists, and even biologists and botanists. And, as the space colonies of the Geospace Region are being established, there are many requests for the psycho-social scientists to make their contributions to space colonization based on their unique perspectives on cultures and societies.

Another set of actors in three-dimensional space is comprised of the robotic explorers and scientists, that perform the traditional functions of the traditional geologists, geographers, physicists and all the other scientists on planet Earth. They do the *field work* in places in outer space that are either too remote or too dangerous for humans to undertake. In doing so, they gather raw data and semi-refined information about the alien atmospheres and surfaces; magnetic fields and plasmas; unknown molecules of gases and solid matter; meteors and dust; solar emissions and eruptions; gravitational fields and... in short, just about anything that affects human exploration and colonization of the Solar System.

A New Geocentric Solar System

The advent of space colonization has required humans to accept – once again – a geocentric view of the Solar System. Cosmology aside, this mental construct of the solar region allows the construction of more useful maps and charts in the colonization of Geospace. Its main advantage is that it realistically depicts the home planet earth as the central point of reference for most human operations in the new human Geospace Region.

A geocentric projection also makes it easier to construct a *georeferenced* grid system for pinpointing and tracking all the bodies – natural and artificial – at any given time and in real-time. Indeed, such a projection of *geospace* as a four-dimensional construct presently facilitates the efforts of the Space Traffic Controllers to locate and track the movement of every object of interest in the region.

Another aspect of colonizing the Solar System has been the surveying and mapping that has gone on throughout the period of space exploration and colonization. Unlike the mapping of the entire Earth, which has taken about 500 years to complete, the process of assimilating sufficient mapping data points from the entire Solar System was completed in less than a century. One reason for this difference is that, prior to the age of hovering balloons, flying aircraft, orbiting satellites and other probing spacecraft, the only way to acquire and catalog data about Earth was by *in situ* surveying of smaller areas of the surface of the planet, and by assembling these various data areas like a jigsaw puzzle.

It was the development of high-altitude photo-reconnaissance aircraft and, later orbiting satellites and spacecraft that enabled humans to view broader areas of the surface of the planet, and to do remote, global and comprehensive surveying and mapping of the Earth's atmosphere, as well as the magnetosphere and the other girdles of charged particles that surround it.

A major stimulus to the continuing and expanding "reconnaissance" of Earth and the rest of the Solar System was the seemingly rounds of weapons races and wars of the 20th and 21st centuries. It would impel humans to do more applied science in space, often through the prism of powerful telescopic systems and the other sensorial systems that were deployed, both on the surface of Earth and in space. Not only were these systems more powerful in terms of their ability to capture the faintest light signal; they also could detect light signals in any area of the electromagnetic spectrum. Then, when the individual telescopes were interconnected by the computers and were integrated into heliospheric networks, their overall contribution to the space colonization effort took a great leap of magnitude.

Thus, one way to define the age of space exploration in the 20th century CE, is in terms of the continuing surveying and mapping of all objects that would become relevant in the later age of space colonization. A product of these mapping efforts has been the development of a large and comprehensive compendium of remote and *in situ* algorithms for developing dynamic map and navigational charts of the entire Solar System... which also has provided the space colonists with the knowledge and skill set that is needed to successfully plan and execute future missions of colonization anywhere in the solar region, and beyond.

Unlike the *hard-copy* maps and charts that were developed to guide the activities of explorers and colonists on during the years prior to the age of computer graphics, space explorers and colonists now have real-time maps and navigational charts that are instantly available as needed. These three-dimensional constructs produce real-time information from a cloud of data, imagery, and information which is continuously updated from a variety of sources. Geospace Traffic Controllers also use real-time electronic projections to constantly track the time-space locations of natural and artificial objects in the same way that Air Traffic Controllers do on Earth with respect to aircraft and weather systems.

Chapter 20

Knowledge-Technology-Skills

The Moon Landing Event of 1969 and the century of human activities in space since then have generated a growing universe of knowledge, technologies, and skills, which are being used in the colonization of the Solar System. Also, during the period of space exploration, many new technologies and skills for carrying out the missions of colonization have been developed by various research and development enterprises; both government-sponsored and non-governmental. Meanwhile, other scientists and engineers have been working to develop optimum trajectories and orbits for colonization; these have been codified and memorialized. Many of these are programmed instructions for establishing optimal orbital insertion points and space rendezvous points; or for performing maneuvers within gravity fields to propel spacecraft. These are available to any space mission planner as "canned programs."

The development of the ways and means for colonizing has been accompanied by a series of rapid advances in the development of artificial enhancements of innate human capabilities. One important example of these is in the technologies of observation. Telescopes, microscopes, spectroscopes and a variety of other observation tools have advanced the ability of human colonists to peer farther and farther into the solar system and the galaxies beyond it; and in phases of the spectra below and above visible light.

Other technologies are being developed for harnessing the radiated energy of the Sun to propel spacecraft and to power their colonies. Huge solar panels are being combined with mirrors and devices to focus and enhance the sunlight to derive as much power as possible from each photon. And, with the use of mirrors, light and energy can be replicated, manipulated and enhanced to serve the purposes of earthling colonists everywhere. Still other technologies are being applied to the harnessing of natural beams of light and making them conduits for transmitting information throughout the heliosphere. And, in the form of artificially cohesive and coherent beams, light can be used both as a tool and a weapon; they can help construct or destruct objects per the wont of the human operator.

K-T-S

The suite of knowledge, technologies and skills needed to prepare and carry out space colonization also has burgeoned during the current period of human history. Each space exploration mission is memorialized with fine detail and nuance that is incorporated into a growing database that can be used to plan new missions, but perhaps most importantly, it is used to create a library of algorithms and techniques which can be used to develop an overall strategy for colonizing the Solar System; to provide what are essentially "off the shelf" software applications to make the development of routine mission plans for colonizations more efficient and, at the same time, provide platforms for customizing missions of colonization in the more remote and problematic regions of the Solar System.

Thus, as humans continue to embark on the missions of colonization, they can draw from a vast and developing "cloud" of experience and know-how… about the areas of operation throughout the Solar System and, the forces, processes, and natural systems within it.

Operational Planning

Mission planning is of the essence in space exploration and colonization, and humans have learned how to plan "customized" space missions, with specific objectives, such as long-distance probe missions to Jupiter, Uranus, Neptune and Pluto; or to study such phenomena as gravity, electromagnetism and plasmas. Other missions are designed to gain practical knowledge about a specific aspect of outer space, such as the Sun's relationship with the rest of the Solar System. And, still others sent to probe the galactic context, within which the Solar System exists and operates.

During the multi-year voyages of spacecraft, which have transited the Main Asteroid Belt, past the Gas Giants of the Jovian Subregion planets, and into outer domain of the dwarf planets and the comets… humans have expanded their knowledge of such phenomena as planetary magnetic spheres, atmospheres, surface landscapes and even the cores of celestial bodies. And, we have also learned much about our own planet in the process. Indeed, each mission report has added to overall human understanding of the whole of the Solar System; and to develop a higher sense of familiarity with our cosmic neighborhood.

And so, in relatively short order, systems have been developed for launching rockets and payloads into low-Earth orbit, initially, and then onto other, more distant orbits throughout the Solar System.

To the same end, many launch facilities and space flight control centers have been established throughout the Geospace Region. The first of these so-called cosmodromes or spaceports were constructed on the surface of the Earth. But as this infrastructure has grown and developed, cosmodromes also have been established on the Moon, some of the near-Earth asteroids, and most recently, on the surface of Mars. Others are strategically located on Lagrange Zones, and some are deployed on orbiting platforms.

Some of these cosmodromes also function as harbors for commercial transportation; they are part of a network of orbiting or surface-based logistical support bases that has been created throughout the heliosphere to provide fuel and provisions to spacecraft engaging in longer or multiple-destination missions.

Chapter 21

Planetary Colonizations

About 1.8 million years ago, the first Pleistocene Human scouts crossed the Red Sea and set foot onto what is now Ethiopia. We can imagine them thinking: "one small step for man; one great leap for humanity". Millions and millions of steps later, Pleistocenes had migrated throughout most of the Eurasian landmass. And later, about 30,000 years ago, Holocenes began a second wave of planetary colonization.

Then, in 1969, modern Holocenes took the first steps on the Earth's moon and initiated what would become the human colonization of the rest of the Solar System. This time, however, humans would make the journeys of colonization on spacecraft that enabled them to transit at speeds measured in thousands of miles per hour: much faster than their walking Pleistocene ancestors; faster than the Holocenes who rode horses and camels across the Eurasian landmass, and who sailed across the seas and oceans, and along rivers; and faster yet than the aviators that fly through the atmosphere.

Whereas, ancient human colonists simply walked under their own power to migrate out of Africa; modern space colonists now use rocket launchers burning about 7 million pounds of chemical fuel, just to escape from Earth's gravity well and to enter a proper orbit in space. And even in outer space, where the gravity field is relatively smooth (except for the gravity wells of the other planets and the less-massive moons and a handful of asteroids), moving from one place to another requires some form of artificial propulsion that must be imported from Earth or harvested from the heliosphere.

The Holocene planetary colonists on Earth advanced migration technology by developing ways and means for transiting the surface of the oceans and other bodies of water; they did so by using the law of buoyancy and the force of the winds and currents to move themselves and their payloads from place to place. Later generations learned how to use stored chemical energy to propel their vessels.

Space colonists adopted many of the elements of the sailing paradigm to transport their space vessels from place to place on the *ocean of gravity* that surrounds the celestial bodies of the solar region. Unlike their sailing ancestors on Earth, however, the *space sailors* would first utilize the stored energy of chemicals and the potential energy of the atom to propel their spacecraft through the interplanetary ocean. Later, they also would develop technologies for harnessing the energy of the Sun and its solar winds to propel spacecraft. And, some use propulsion systems that use solar energy and chemical energy hybrid technologies, and the energy derived from decaying atoms.

The interplanetary medium is like the oceans of Earth in many ways. Its relationship with the celestial bodies is reminiscent of the relationship between the oceans and the continents. However, unlike the oceans, which are comprised of liquid water (and some ice), the materials within the space *ocean* include neutral (non-charged) hydrogen and of plasma gasses that are essentially fields of electrically charged particles that have been emitted by the Sun, cosmic rays.

On the other hand, like the oceans, the interplanetary medium is in constant motion. Contributing to this dynamic are the solar winds that are continuously emitted by the Sun, the emissions from the stars of the galaxies, the currents of charged subatomic particles, and even the dust that derives from continuing cosmic tectonic activities. Like the ocean waves, these cosmic and solar emissions propagate through the interplanetary medium until they encounter a planet or some other celestial body. The solar winds ordinarily would barely register on an anemometer on Earth, but what they lack in force, they make up in persistence and constancy. Indeed, they are strong enough to impart a slow, but persistent and cumulative force on a spacecraft, thereby enabling it to "sail" the space ocean. Moving about in the interplanetary space is also analogous to "surfboarding" the waves and troughs of the waters of the ocean. The topography of the gravitational continuum acts on a spacecraft the same way that the swells and troughs do on a seagoing vessel.

Chapter 22

Navigating The Gravity Ocean

During the century or more of space exploration and colonization, humans have learned that the key to moving about in the solar system is to navigate the orbits: in much the same way that seafarers use the ocean currents on Earth. Orbits are manifestations of the *gravitational negotiations* that constantly go on between the Sun and the other celestial bodies; and between the latter.

Each of these gravitational negotiations occur within the parameters of gravitational fields, which are the product of the interactions between the two main variables of mass and distance, although external variables like intrusions by asteroids and comets, or unusual perturbations caused by solar eruptions also are operating variables.

During the decades of human operations in space, mission planners, astronauts and engineers, and many other space specialists have learned many of the secrets of the various fields (natural region) that occur throughout the heliosphere, which is itself a field that consists of the flux of the solar wind. Other significant fields within the heliosphere include the magnetospheres of many of the planets and other massive celestial bodies. There are the also the other fluxes of plasma which create fields of charged particles at various places in the solar region.

And, as was the case with Pleistocene colonists with respect to fire... the Holocene earthlings have also gained a sufficient degree of *familiarity* with the force of gravity to begin developing ways and means (knowledge + technology + skills) to begin harnessing the properties of gravity and the gravitational field to conduct operations within the Solar System.

As they have gained this deeper knowledge of gravity and its various permutations of effect, humans realize that they must respect the rules of the gravitational field in which they are operating. However, they also have learned that these parameters can be *negotiated* through the application of technologies and techniques. Artificial propulsion and throttling are examples of such measures; another is the utilization of the angular energy of rotating massive bodies to add momentum to spacecraft is an example of negotiating with gravity fields.

The laws governing the relevant gravitational field also play an important role in travel and transportation within the solar system. The controlling variables are always related to mass and distance, but in the case of moving objects, like asteroids and spacecraft, *distance* is always changing, always relative to the velocity of the moving object.

Distance becomes the alter expression of time in space travel: time-distance. Thus, spacecraft are traveling at high speeds that effectively shorten the distance between the point of origin and the destination. On the other hand, the effective distance is increased by perturbances that act like headwinds in atmospheric flying. Distance multipliers and hindrances occur with internal changes in the gravitational field or from externals, such as the nearby transit of asteroids or comets along the trajectory of the spacecraft.

Principles of Travel and Transport in Outer Space

> Every spacecraft-payload unit (SPU) can be thought of as being a *cultural celestial body*... It, therefore, has the property of mass and must navigate the gravitational ocean in the same manner as natural celestial bodies...
> The greater the mass of the SPU, the greater the demand for propulsion power...

SPU mass is determined by the sum of the mass of the payload + spacecraft + propulsion subsystem...

Mass determines the amount of external force that is necessary to move an SPU point to point. This destabilizing force and the resultant momentum it imposes on a spacecraft is applied in the form of released "stored energy," as part of an artificial propulsion system...

Therefore, the key to efficient travel and transportation within the gravitational field is to minimize the mass that is to be moved... and/or to maximize the potential energy inherent in the propulsion system.

Efficiency of movement through a gravitational field can be maximized by using any combination of lighter materials; more efficacious propulsion systems; and micro-miniaturization of payloads...

Applied Space Navigation

The developing principles of space navigation have been applied in planning such missions as finding the optimum pathway for transiting through the Main Asteroid Belt, or a longer mission to Mars and Jupiter. In each case, in which a spacecraft is travelling through the combined gravitational effects of a variety of relevant massive bodies, the optimum mission plan may involve several navigational maneuvers, including the so-called "slingshot" maneuver, which uses the angular momentum from one or more celestial bodies to enhance momentum.

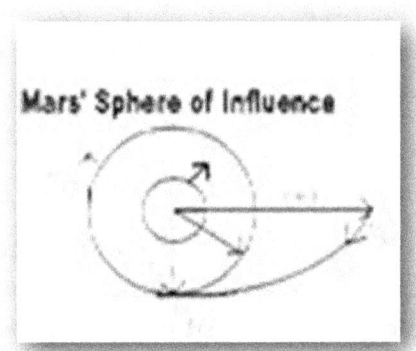

Credit: Massachusetts Institute of Technology (web.mit.edu)

Thus, choice of orbits and trajectories for a given mission is fundamentally driven by the mission objective(s). Within these broad parameters, however, the choice of orbits and transit trajectories is constrained by the relative alignment of the celestial and cultural bodies (e.g., space colonies) that lie along the point of origin and the terminal destination. The individual rotations and the movements of the relevant planetary bodies along their cyclic orbit paths also must be taken into consideration in planning a space mission, and in selecting a site for a space colony or other type of space installation in orbit, for that matter.

Traveling along an orbit, like any long sea voyage on Earth, involves the conduct of a continuing series of negotiations between the spacecraft and the gravitational fields it transits. Even a relatively low-mass body such as a space probe, an orbiting satellite, or a space shuttle-tugboat configuration represents an object with significant mass as it goes forward, especially as the number of relevant natural bodies increases. Thus, even as the transiting spacecraft tries to maintain optimum momentum along its orbital path, the gravitational effects of other bodies in space can work to destabilize it. So, there is a continuous tug-of-war between the one object wanting to maintain course and others wanting to change its course.

So, the dynamics of spaceflight involve two main forces that operate on the transiting spacecraft: (1) the net product of the competing gravitational attractive and repulsive forces that are generated by massive bodies or magnetic fields and, (2) the net velocity of the spacecraft, which is a function of the force from the artificial thrust provided by from the integral propulsion system and externals like the angular momentum received from nearby massive bodies. One of the most common of the latter is the so-called "slingshot effect."

The Interplanetary Gravitational Field

Like the myriad of waves, currents and other disturbances and perturbations on and under the surface of the earth's oceans, there are perturbations within the gravitational field of the heliosphere, mainly caused by the meandering orbits of the planets and moons, or by the intrusions from starbursts and solar storms, or by collisions and near-misses caused by transiting asteroids and comets.

As earthlings began to explore interplanetary space *in situ*, they realized that the moving-about the three-dimensional space is analogous to moving about the three-dimensional space of the Earth's atmosphere, and within bodies of surface water on the home planet. Hence, as the spacecraft have continued to travel throughout interplanetary space, they have experienced gravity in ways that were quite different than on the surface and in the atmosphere of Earth.

How the humans experience of gravity in outer space also differs from that of planetary atmospheres (where they exist) throughout the Solar System. And, when *silicon-based (SB)* humans began to explore the near-surface gaseous 'oceans' of the Gas Giant planets, they encountered evidence of the same gravity-effects that are found in the liquid bodies on Earth.

Consider that earthlings have evolved to operate at optimum levels within most of the gravitational environments (GE) on the surface and within the lower strata of Earth's atmosphere, without the need for any external technology. In both the upper atmosphere and the depths of the oceans, the effects of gravity are well-understood, and the terms of the negotiations are well-known and carried out by the application of tested technologies. Increasingly, the same can be said for the state of navigation in outer space. As humans log more hours there, and on the celestial bodies, they have determined the inputs of technologies that are necessary to reach a satisfactory accommodation with the various gravitational environments.

All these variables are all considered and evaluated by planners prior to a mission; and by space navigators throughout the course of the mission. These calculations consider the predicted locations of relevant bodies prior to the mission and through continuous *in situ* calculations based on analog data received by sensor-actuator systems, processed by computers, and outputted to the controllers – all in real-time.

Then there are the ubiquitous micro-effects from perturbances that affect the transit of spacecraft through a *gravitational field*. These waves and currents are a net product of the competing attractive and repulsive actions of the massive bodies of the Solar System – as well as the occasional eruptions of the Sun and the galactic stars. The latter not significant in determining the movement of spacecraft on relatively short-distance missions, but they do affect long-distance travel to the outer reaches of the heliosphere; out to the hinterlands of the Kuiper Belt and to the heliopause which marks the end zone of the Sun's domain, and the beginning of the jurisdiction of the Milky Way Galaxy.

Gravity Fields

Astronomers and other space scientists have developed a digital map of the Solar System based on a body's distance from the Sun and its mass, orbital profile, and so forth. This has served the primary purpose of providing a logical context for developing catalogs and maps of *masses and distances*. It serves well the exploration and colonization of the solar region.

One of the most important factors which determine the relative distances between the Sun and the planets is the absolute and relative mass of the latter; the relative distance of less massive celestial objects from the Sun is also affected by the mass of their nearest neighbors. A corollary of this rule is that the net relative distance of a spacecraft (including space colonies) from the Sun is determined by the combined gravitational attraction of the Sun and all the other massive objects within the spacecraft's *neighborhood*. The other relevant variable in the calculus is the distance between said celestial body and the transient spacecraft or natural body.

This brings up the basic problem in navigating between two points in space, which is: (1) the interplay of competing gravitational attraction and repulsion forces throughout the three-dimensional *gravitational ocean* within the solar system, and (2) relative displacement of massive bodies of the surrounding gravity field.

The bodies with the greatest concentration of internal matter in the heliosphere are the Sun and Jupiter. These produce the deepest *depressions* in the gravitational field of the solar region. Other planets cause more shallow depressions in their area of the medium, and moons and some asteroids have lesser and more localized depressions. In any case, all bodies that contain mass, even comets and spacecraft create some warping of the interplanetary gravitational field.

Another way to characterize the depressions in the gravity field of a massive bodies is in terms of the contours of the *gravitational wells* within it. The practical effect of these extreme sinks is that a spacecraft is required to apply an extra degree of propulsion to escape from the well; that is, the practical effect of gravitational wells is to create an attractive force which hinders the escape from occurring. Or more generally, the only way to escape the attractive clutches of the massive body is to exert some external force to propel the less massive body onto the *surface* surrounding the gravity well by the application of external *artificial propulsion*.

Once an escaping spacecraft body breaks free (reaches escape velocity) it must enter an insertion point along an orbital path and then reach a minimum speed to maintain the orbit around a massive body. Thereafter, a space navigator *sails* gravitational "ocean" in basically the same manner as aviators do on Earth's oceans, by utilizing the "lift" from centripetal forces and the "thrust" from artificial propulsion.

In some ways, transiting through interplanetary space is more straightforward than flying through the Earth's atmosphere. In the near-vacuum of space, there are only minor turbulences caused by the gravitational influences of bodies that occur along the trajectory of the spacecraft. Otherwise, there is practically no drag or headwinds involved, except perhaps from the relatively docile *solar winds*.

However, interplanetary space is far from "empty." It contains charged particles from a variety of sources, including cosmic and solar radiation, plasmas and all manner of electromagnetic energies. There is all sort of matter too – from gas and dust, to rubble from planetary building and collisions… to streaking comets and asteroids… to planets of all sizes and their satellite bodies… And now, there are artificial bodies, space junk, and even stray nuts and bolts from structures that have been constructed by human space explorers and colonists.

In practical terms, the basic calculus of movement through interplanetary space consists of the relation between propulsion (if any) and the massiveness of the spacecraft itself. This holds true for the smallest orbiting artificial satellite and the largest orbiting space colony. From this basic equation, space colonists have derived a suite of technologies aimed at minimizing the mass of objects in space, and of maximizing the efficiencies of propulsion as much as possible; to this end, the Enlightened Minds of the Geospace Region continue to search for the lowest common denominators of matter and energy, to construct objects with the greatest efficiencies of size and mass for operating in space.

"Dark Space"

During the middle of the Holocene, several Enlightened Minds have received continuing flashes of intuitive cognitions about the still empirically-unknown "truths" of the cosmos – what today be is described metaphorically as *dark matter* and *dark energy*. From these illuminations of the mind, they have continued to develop, in their mind's eye, non-sensory images of the elementary matter and energy: natural phenomena that are not visible to the human eye; that cannot be heard by the ear... nor felt by the skin.

On the other hand, the human organism – since the time of the GHAE – has appreciated some properties of *dark phenomena* through derivative *analog* data that is captured by the organic senses, including those of the hearing, smelling, tasting, touching, and seeing. And, although the unaided human eye cannot appreciate infrared energy directly, the body can *feel* the derivative heat and the brain can produce imagery through analog logic.

It was the Enlightened Minds that first peered into the secrets of electromagnetism. Thus, the secrets of energy were intuited from data points occurring in nature, such as amber cloths that emitted strange sensations when rubbed, or the crackling stream of light and sound of lightening, and the fact that the *flow* of the lightning bolt across the skies seemed to target certain points on the surface of the planet.

They also intuited the relationship of the atom's nucleus and its orbiting electrons with electromagnetic energy, and to the energy derived from the fission or fusion of nucleus and of the energy that is released every time its electrons leave the atom and migrate to other atoms; and how the atoms relate to the living cells, which contain programmed instructions for constructing all living matter, and photons which transmit the energies that guide the form and behavior of all states of matter.

Most significantly, the Enlightened Minds learned the power of analog logic to elucidate the secrets of *dark* natural phenomena. Thus, for example, they utilized their familiarity with the flows of rivers to intuitively comprehend the fluxes of charged particles; as analog for understanding the workings of electromagnetic energy as a dualism of waves of energy and their charged particles.

Chapter 23

Energy is Matter is Mass

The relationship between mass and energy is contained in the equation:

$$E = mc^2$$

--- ALBERT EINSTEIN

This pithy but powerful statement of the relationship between mass and energy was a crowning achievement of Holocene humans; and more particularly, of a mutant mind that initially arrived at this understanding mainly through the power of intuition. Later, other Enlightened Minds of the 20th century delved into the nature of each of these fundamental phenomena at the macro and micro level.

The 20th century saw a growing understanding of the atom and each of its grand constituents: the nucleus and the electron. A nano-scale look at the atom itself revealed a nucleus which contains a varying complement of neutrons and protons, and which is surrounded by a varying number of active electrons. There also was an inherent and alternately organizing/disorganizing effect from the energy that permeated this basic component of matter in the universe.

This inherent relationship between mass and energy, the speed of light, and the tremendous kinetic energy that is released in every transubstantiation of energy to mass and vice-versa, was initially domesticated by Holocenes to create atomic explosions in warfare. However, during the next few decades, humans also learned how to harness the great potential energy that is found in the nucleus of the atom for peaceful purposes, such as in rocket propulsion.

The essential lesson of $E = mc^2$ is that energy can be derived from mass; that mass that is moving at the speed of light-squared is equivalent to energy… and that, therefore, energy is mass in a dynamic form. This truth about mass-energy has been used in many applications by humans as they set about to colonize space… indeed, it is axiomatic that *energy bodies,* such as plasma fields, electric and magnetic fields, cosmic and solar radiation currents affect space operations in much the same way as the massive celestial matter; and that energy bodies also introduce turbulences in the interplanetary gravitational field.

Holocene humans discovered many of the secrets of energy as early as about 4,000 years ago, when the human mind had matured as an entity of higher and further knowledge. With these new powers of *a priori* knowledge, humans now could apply abstract cognition and logical analysis in the processing of cosmic data. This alternate trajectory of knowledge, which extends backward and forward along the time-space continuum, is the source of the great insights that Enlightened Minds have acquired, at various times and places, to further the colonization of the entire Solar System.

The powers of the mind have also been used to elucidate the secrets of the *what* and the *how,* which they then have transmitted to the human brain, through the process that has come to be known as the *scientific method;* that is: *c*areful examination of data, a practical refutation of theological and other dogmas as being axiomatic; the *clever* development of analog models; the direct testing of hypotheses drawn from these models; a willingness to accept the results of the testing even when they contradict the models of presumed or *a priori* notions of realities; and the persistent trial and error iterations that eventually led to the next level of familiarity of natural forces and entities (energies and masses).

Walter Gomez

BY THE TIME OF ACTUAL SPACE COLONIZATION, HUMANS HAD BECOME "FAMILIAR" WITH MANY OF THE ELECTROMAGNETIC ENTITIES IN SPACE...
SUCH AS:
Electrical and magnetic flows and fields
Plasma flows and fields
Charged particle flows and fields
Wave/Particle flows and fields
Cosmic and solar radiation of charged particles
Solar storms
Atomic decay radiation – alpha, beta, gamma rays

Then there is the myriad of charged particles and electromagnetic flows, which mimic the currents and winds of the atmosphere, and the currents of the Earth's oceans. These phenomena are now relevant to the human experience because of the way they affect the transit of bodies through the a medium. As a result, are now well-known and are portrayed on navigational charts of the interplanetary medium that resemble those of the oceans on Earth. Thus, for example solar eruptions and the consequent acute propagation of waves of charged particles are presented on electronic displays with the same imagery as are the tsunamis on Earth.

So, the macro expressions of charged phenomena is more obvious to the natural senses, but other manifestations of electromagnetic phenomena can only be "seen" with artificial sensors, either because they are too small or they are too opaque to be detected by the human eye… or because they occur and travel in wavelengths of light that occur along wavelengths that are either shorter or longer than those of visible light. Still other *dark phenomena* are still only known indirectly by analog inference – as in the case of the slight perturbations of planets – thought to be the result of the gravitational effects of dark phenomena.

Highly-charged and penetrating electromagnetic radiation in the higher-frequency (shorter-wave) gamma rays are clearly and presently damaging and even lethal to the cells of humans. On the other hand, electromagnetic radiation in the lower-frequency phases: microwave, radio and ultra-violet are generally less harmful to humans, and are even useful as a tool for space colonists. As an example, in the form of beams, they can be used to transport bits of information and pixels of imagery – at virtually the speed of light.

Consider that space colonists need to be able to *foresee* what is ahead, as they venture out to the places in space where they want to establish orbiting or surface-based colonies. With the use of the electromagnetic spectrum tool-kit and associated skill-sets, they can *actively* transmit radar, infrared or even visible-light signals and then detect and analyze any reflective signals to create a virtual image of what lies *over the horizon*, as it were. In some cases, the receiving and retransmitting point will be a natural body whose albedo (reflective properties) are known – or it will be an artificial *beacon*, which is programmed to retransmit reflective signals in a programmed frequency.

In a similar vein, there are communication relay nodes, which have been preprogrammed to operate at certain frequencies, within a network for all manner of communications. These have been deployed throughout the Solar System to operate like the servers of a global computer network, as they coordinate the flows of information throughout the network of colonies and other bases.

Chapter 24

"Weather"

Weather is both a natural phenomenon and a human construct. It finds meaning only in the context of location and in its effects on humans and their machines. There is "planetary weather" and "space weather." Each is different in terms of inherent properties and in its effects on humans; and it is humans who everywhere give meaning to the *elements* of weather, including temperature, precipitation, wind and ambient pressure.

Prior to the beginning of human colonization outside Earth, the notion of *weather* had been applied only to the atmosphere, the surface, and the subsurface of Earth. These were the places where humans lived in the days before the deployment of the International Space Station. Since then, however, ambient pressure and other aspects of gravitational forces have been felt and understood differently than they were on Earth.

Consider that when ancient humans set out to colonize planet Earth, they encountered the virtually the same weather conditions that they had experienced throughout their incubation period in Africa – at least within the Tropic of Cancer Region. It was only when they ventured outside their traditional comfort zone, into the higher latitudes of the northern hemisphere that they were confronted with colder temperatures which required technological responses.

Meanwhile, the weather within the TCR remained relatively unchanged throughout the first 1.7 million years of the Pleistocene Period. There were short-term variations in temperatures and in precipitation; but within the experience of any given generation of humans or within the lifetime an individual human, the existing pattern of weather seemed "normal." So, over the short-count period of common experience of each generation, Pleistocene colonists barely took note of the alternating changes in the magnitude of the ice domains of the higher latitudes.

The early humans only felt changes in temperature throughout the course of the day and night. They also recognized that highlands tended to be cooler than lower-lying valleys. But overall, temperatures within their domain on Earth tended to remain within a tolerable range of comfort; one that could be "managed" by seeking shade during the day, or wearing animal skins when temperatures were cooler. And, shelters and fires could be utilized to sort of *air condition* living spaces within natural or constructed shelters.

Changes in the patterns of precipitation were the main aspect of weather for Pleistocene humans; they were the controlling variable which determined where they would find hunting-grounds, with adequate and accessible sources of water; water that would sustain plants and animals, both of which were sources of food for humans.

Consider that the Pleistocene colonists did not manage the inputs to their food production systems; that would not happen until the emergence of the Holocene colonists. But at least the Pleistocene colonists could follow the lead of the grazing animals as part of the strategy of "following the water." However, following the water strategy would become increasingly problematic during the latter periods of the Pleistocene Period. In fact, the current Ice Age was ending, the ice domain was definitively melting and a great volume of water was being added to the oceans and the atmosphere. "Times and places were changing."

Cosmic changes in the revolution and rotation of planet Earth would be the catalyst that would cause the consequent renovation of most of the planet's surface. It had happened before, but this time, there were human populations involved. The very coastal lands on which they historically used in their migrations, and from which they drew their sustenance, were slowly disappearing. The traditional tactics for survival were being disrupted and whole populations were becoming refugees, rather than proactive colonists. A major factor in causing this diaspora was the fact that water was only useful to hunting-gathering humans when it collected on the surface, so that it could be used for drinking, cooking and other purposes. This meant that they had to constantly react to changes in the surface waters in the manner of the herding animals.

Indeed, one of the positive consequences of the disorganization of traditional sources of water within the Tropic of Cancer Region during the *time of troubles* was that it impelled the Late Pleistocenes and the Early Holocenes to develop new sources of water below the surface of the planet.

Even before the remaking of the map of surface waters caused by the tectonic changes of the waterscape and landscape, which began about 1.7 million years ago, some Pleistocene hunter-gatherers – especially those that wandered along the coast of the oceans, seas and inland lakes – must have discovered that simply by digging a foot or so below the surface – fresh water often filled in the hole. This knowledge eventually would have propagated to other groups of Pleistocene migrants who would have carried this knowledge as they established central places throughout the Tropic of Cancer Region. Indeed, as early as 1.7 million years ago, there is forensic evidence that humans were tapping artesian wells in search for a steady supply of fresh water.

Cultural Response

So, even during the first period of diaspora, some Pleistocene Enhanced Minds already were planning a cultural response to the new realities of their world. They had, by then, developed a significant body of *cultural wisdom* and a shared *mental map* of a new Tropic of Cancer Region, and intuited that a major change in their world was coming. And, equipped with this shared wisdom and mental maps, many Pleistocene colonists of the time began to focus their attention on developing stationary colonies on Earth.

And so, rather than simply *living off the land* by foraging and finding targets of opportunity, some groups of these so-called "Archaic Homo sapiens" sought out places that contained unusually high concentrations of the most vital resources for the survival of the colonies. The most favored places were those with *accessible* water, of course. Not only for drinking, but also because water also tended to be associated with plant and animal life, food and fiber would be available as well.

So, even amidst the diaspora caused by the renovation of the Earth's surface, some Pleistocenes were trying to make a successful transition, from a mobile hunting-gathering existence, to one based on settled places; where they could make a living by cultivation, rather than extraction of resources.

ONE MUST STAY IN A PLACE LONG ENOUGH TO EXPERIENCE SHORT AND LONG-COUNT CYCLES OF ATMOSPHERIC EVENTS TO 'GET TO KNOW' THE WEATHER THERE…

The Holocene Humans decided to gain some control over their food resources; to lessen their *food anxiety* by ensuring a more reliable production of the essential nutrition that is needed to survive and thrive in each colony. They, therefore, decided to remain "in place" and to exploit the resources of a given place more intensively. This approach led to the development of organized and purposeful cultivation of both plants and animals for making food and fiber. And, the decision to remain in place also generated the need to study more closely the life-cycle and properties of the natural resources of the local place, and to study the environmental phenomena that affected the plants and animals which they domesticated.

A new existential paradigm was emerging. Because of the decision to stay-in-place to make a living, Holocene colonists found that they would have to work harder and smarter to acquire food and fiber, and that the payoff

would be a lessening of the pressure to acquire food from the wild, every day. But this meant that a different suite tools and skills would be needed to successfully carry out this new strategy for colonizing a planet. And, it also meant that humans would have to learn a great deal more about the life-cycle of plants and the local ecological system that determined their nutrition potential. And, it meant that some form of record-keeping would have to be developed.

One derivative of this form of intensive food-production was that the human construct of *weather* came into clearer focus. "Weather" now became an omnipresent element in human existence in any given place on Earth. Its secrets would have to be discovered through the process of persistent and disciplined observation; careful and faithful recording of these observations; and the utilization of this knowledge to create harmonious agricultural systems within each colony.

So, there began a continuing effort to learn about and even to predict when the rains occurred both locally and throughout the regional watershed. The Holocenes became aware of *weather*, as it was occurring, and in within an expanding context of patterns. Some Enlightened Humans even began to realize that local weather regimes (climates) could be affected by weather in faraway places.

Even well into the Holocene Period, the idea that external phenomena could affect local rainfall patterns was often ascribed to the behavior of gods or other such entities. But despite this, humans continued to learn as much as possible about local rain patterns through persistent empirical observation and logical thought – while also seeking ways to encourage the assistance of the rain gods.

The Age of Industry

Fast forward to the 19th century CE. The central places of industry are becoming the dominant nodes in the global clusters of human populations. It is also the time when the machines are rising, as they are being powered by engines fueled by fossil remains of ancient plants and animals. It is the age of science, during which humans are more becoming more secure in their reliance on human powers of empirical observation and the scientific method as a means of solving the existential challenges of colonizing the planet.

There also was a trend towards using machines to do more of the work that had traditionally been done by humans and animals at the turn of the early 20th century everywhere on Earth, even in the "developing regions" where the cultivation of plants and husbandry of animals for food continued well into the Space Age. So, the age of machines was trending along the same trajectory of migration of people from the rural provinces to the urban centers. Machines and fossil-fuels also begin to change the way humans, goods and services were being transported within and between central places. Ships increasingly were being powered by engines using fossil fuel, instead of sails and wind power. Railroad engines, powered by wood, coal and diesel were replacing horse-driven vehicles. Electricity became *familiar* as it was harnessed to power engines, to light up the night, and to convey information via wires and cables at nearly the speed of light.

Enter the 20th century, and humans are exploring the atmosphere and flying through this medium in powered aircraft. Rockets use the potential energy of explosive chemicals to extend the exploration of the skies beyond the gravitational attraction of Earth's core, to access the *outer space* that lies between the planets and other celestial bodies of the Solar System.

And now, in the 21st century, modern humans have developed nodes and constellations of orbiting space colonies – first in low-orbit around Earth, then in orbit around the Moon. Other constellations are being developed within Lagrange Zones which orbit the Sun.

Throughout all these changes in the way of living and in the ways of making a living… weather is revealed as being more complex as a natural phenomenon, and in the ways in which it affects human efforts to colonize the Solar System.

Extraterrestrial Weather

By the time the first probes were sent into the stratosphere, the observation and study of weather already had developed into the comprehensive and sophisticated science of planetary meteorology. Humans had achieved the ability to not only track but predict the movement of weather systems on Earth; they had become *familiar* with the various elements, such as wind, precipitation and temperature that fall under that rubric.

However, when humans began operating in outer space —outside the field of atmosphere — they entered a new ambience of virtual "non-weather" in which there is no discernable wind and no precipitation in the earthly sense. The only element of traditional weather they experienced was temperature. Suddenly, the definition of "weather" changed; in outer space; it now also referred to the flows and fields of charged particles. However, on the planets and their moons, humans once again encountered "precipitation" of ammonia or methane, sometimes mixed with water. And, there would be other variations of weather that is analogous to the weather on Earth.

As it happened, the Cosmic Mind had already transmitted to humans the powers of acquiring *a priori* knowledge of the natures of *weather* throughout our Solar System, even before the first space probe had been sent to explore the transitional zone between earth's atmosphere and outer space. And, with illumination from the Cosmic Mind, humans also were fortified the knowledge and power of logic to develop solutions to the challenges of human survival in any given weather regime.

One of the unique capabilities of the human mind is its ability to recognize *analogs* between past, present and future realities without any interaction with the senses of the human brain. So, even before humans had begun operations in space, they already had a wealth of *a priori* knowledge of the weather in outer space and on the planets and other celestial bodies that lie outside the Earth's atmosphere. So, today, through the power of *a priori* cognition and analog conceptualization, space colonists can continue to develop theories and models of the weather throughout the Solar System — and to develop technologies for living and working in the new weather paradigm.

In outer space, precipitation and humidity, as well as winds and sandstorms and geysers are still elements of the weather, but only on certain celestial bodies. On these bodies, space scientists and prospective space colonists seek to learn about their weather regimes by comparing them to places on Earth which are deemed to have analogous weather conditions. These include the weather regimes in places like the Atacama Desert of South America or the Gobi Desert in Asia or in Hawaii's volcanic regimes.

So, the interplanetary medium contains particles of dust and microscopic particles that are left-over from the initial construction of the Solar System, and the continuing tectonics that are continuing today. Some dust particles borne by the solar wind; others are expelled by asteroids and comets as they transit from the outer regions of the heliosphere towards the Sun; and planets also expel dust particles, because of volcanic eruptions or impacts by external bodies.

The virtual absence of breathable oxygen in space also has an analog on the highest elevations of the mountains and in the depths of the oceans. Humans who have climbed to the summit of Mount Everest have experienced what it is like to breathe air that is devoid of adequate concentrations of oxygen atoms. The same phenomenon has been experienced by balloonists and by airplane pilots in the days before the advent of bottled oxygen and pressurized-oxygen cabins. But sometimes the analog of an element of weather is most powerfully manifested in its opposite expression; an example of this phenomenon is the ambient pressure that surrounds humans on Earth; it is this omnipresent, but variable, external atmospheric pressure that maintains the integrity of the internal organs within the outer skin of the human body. It functions well enough at lower altitudes of the atmosphere, but in the higher altitudes of the stratosphere and in the deeper levels of the oceans, the ambient pressure is either too low or too high to maintain the vital integrity of the human body.

The Solar Wind

The solar wind in outer space also has its analogs on Earth; it differs from the atmospheric winds in kind, but its behavior is similar. Both flow through a medium at various intensities and both affect the medium through which they flow. However, the solar wind differs from the atmospheric wind in that it consists of microscopic charged particles that flow through the heliosphere in accordance with the laws of electromagnetism. Another critical difference between atmospheric and solar winds lies in their effect on the human organism: solar winds carry charged particles that can penetrate the cells of the human skin and the vital organs within it.

Some flows are more energetic; have greater impetus than others and can, therefore transit greater distances. The higher-energy particles travel within shorter waves; and they flow past a given point with greater frequency than longer-wave fluxes. Higher energy waves have more powers of penetration and therefore are more dangerous to human cells (and their electronic systems).

Because energy move at nearly the speed of light as they propagate throughout space, they are useful as reliable carriers of *information* of various kinds. But their speed also makes them dangerous to humans and to their structures and machines because they can penetrate and damage both organic and non-organic materials.

Planetary Weather

Geospace colonists eventually would internalize that "precipitation" could consist of both the cosmic and subatomic particles, and of the atmospheric phenomena they experienced on Earth. The settled on the rubric of "cosmic precipitation" to describe the former and "planetary precipitation" for the latter.

The most prominent sources of cosmic precipitation in the Solar System are the galactic stars and the Sun, as well as the ubiquitous cosmic cloud of charged and neutral particles leftover from the Big Bang Event. And then there is the precipitation of charged particles that are emitted by the constant mutation and decay of atoms of many elements.

Planetary precipitation *can* occur on the planets and moons which have developed atmospheres. In those places, precipitation occurs, but it may not be in the form of water; it may be as methane or some volatile.

Winds occur on many celestial bodies and temperatures rise and fall, as per the nature of the body's exposure regime to the Sun's rays. Volcanic eruptions and flows, and other pressures from the planet's interior activities can also raise temperatures on the surface and atmosphere of a celestial body. Celestial bodies also experience winds and fronts like those on Earth.

Cosmic Radiation

Galactic Cosmic Radiation (GCR) consists of atomic nuclei that originate near black holes or neutron stars, during supernova explosions and in the hot aftermaths of these events. Others are produced by the radioactive decay of atoms. Smatterings of neutrinos and gamma ray photons are also emitted. All these high-energy, high-velocity emissions produce swarms of particles that travel at ultra-fast speeds; especially the *tsunamis* of gamma rays.

The propagation rate from the stars also depends on the initial charge of the particles that are being emitted. On the other hand, the penetrability of these charged particles through a medium is attenuated by the distance from the star or by the shielding profiles presented by bodies of matter or energy they encounter.

It should be noted that the galactic cosmic radiations undergo significant changes during their journey through the galaxies and the heliosphere. Even before entering the Solar System, galactic fluxes will have experienced many transubstantiations and mutations which will reform their character. And, once within the Solar System, these fluxes will continue to experience changes as they interact with magnetic and electrical forces and fields, as well those of gravity and the strong and weak forces. Thus, as radiations of charged particle cross the heliopause (the frontier zone between interstellar and interplanetary space) they encounter flows and fields of solar radiations, which also will reform them. All these forms of electromagnetic and nuclear radiations will then interact and combine with other fields and fluxes of charged particles; both positively and negatively charged elements will interact with each other and with electrons that are still attached to the nucleus of atoms, to form a variety of molecules and, thus, creating many iterations of the traditional elements of the periodic table.

In any regard, the law of conservation of energy will be observed – even as the release and emission of energies continues along a spectrum of intertwined electricity and magnetism. And, all the while, atoms will experience many changes and reconfigurations during their journey through the galaxies and the Solar System. Within the atoms: as electrons shift from one orbit to another; or leave the orbital structure of a given nucleus – energy will be emitted. Atoms whose nuclear integrity is compromised also will release energies with differing values of penetrability and at velocities nearly as great as that of light. Ultimately, though, it is the potential penetrability of the particle, regardless of its progeny, that is relevant to space colonists. Penetrability implies the capability to degrade or destroy the integrity of atoms of inorganic materials or of cells of living beings.

The Sun's emissions are initially generated from its core through the process of nuclear fusion. Then, via the process of convection and interaction, some of the radiated energy will eventually reach its outer surface where, within the corona the convection and mutation processes will continue, until finally, the net emissions are expelled forcefully and suddenly; or propagated gently but surely, onto the heliosphere. In either case, some of these charged particles will arrive at the magnetosphere… the atmosphere… or even at the surface of some celestial bodies.

The Sun also projects a magnetic field that interacts with other fields and fluxes throughout the Solar System. This magnetic field varies in magnitude through time and space, and its most prominent manifestation is the so-called sunspot, which appear as a visible, dark patch on the Sun's photosphere. Sunspots are relatively cooler than the surrounding photosphere and therefore appear darker than the surrounding surface.

Sun spots are only one of several indicators of cyclic variations in the strength of solar radiation and the solar magnetic field. Consider that the sun also produces a variety of other space *weather* phenomena, including the Coronal Mass Ejections (CME), solar flares, and the high-speed streams of charged particles and plasmas which are carried by the solar winds.

The most common conveyor of solar radiation is the solar wind. It carries with it cascades of charged and neutral subatomic particles, consisting mostly of electrons and protons. The solar winds vary in temperature and speed as they interact with cosmic currents, other plasmas and magnetic fields during their journey through the heliosphere. And, they convey electric and magnetic fields, plasmas and other charged phenomena, which vary in their ability to penetrate various magnetic fields and various materials including, soils, rocks and certain metals.

Geospace 2060

THE PRACTICAL UNDERSTANDING OF COSMIC AND SOLAR RADIATION HAS PROVEN TO BE AS SIGNIFICANT TO SPACE COLONISTS… AS FAMILIARITY WITH FIRE HAD BEEN TO THEIR PLEISTOCENE ANCESTORS MILLIONS OF GENERATIONS EARLIER…

Consider that even mild solar flares and slight increases in solar wind radiation levels can interfere with the functioning of orbiting satellites and orbiting space colonies. More acute gusts of solar winds, whether they are produced by imploding stars or solar eruptions and storms, can cause much more widespread and more destructive effects on their electronics and telecommunication systems in space, and even within the protective umbrella of magnetic fields and thick atmospheres. And, as the proliferation of wireless communications increases, the impact of solar flares and peaks in the solar wind will only increase.

Another dimension of the problems that are caused by solar flares and other eruptions of charged particles is the result of the sheer increase in the number of human spacecraft in the interplanetary medium and in the number of surface-based installations on planets, moons, asteroids and even some comets.

Sunlight

Unlike other kinds of cosmic radiation, the radiation of from the Sun is more beneficial than harmful on balance. Its deleterious effects are more easily managed and, at the same time, its beneficial properties can be enhanced, controlled and otherwise managed for human purposes. The key to managing both negative and positive effects is to control the rate and duration of the sun's rays on a specific place of human activity.

An important factor in determining the rate of sunlight on both natural and artificial celestial bodies is their pattern of rotation and their orbital patterns with respect to the Sun. Therefore, colonies that are deployed on the surface of natural celestial bodies must take into consideration the diurnal rotation cycle of the body to manage the amount and duration of solar light. The available methods for managing sunlight on a celestial body have been to: deploy the colonies at several strategic points on the planet to maximize the diurnal quantity of insolation; to deploy artificial solar radiation collection stations; and to amplify and focus the solar light through a network of mirrors and similar reflectors to transmit sunlight when and where it is most needed.

On the other hand, orbiting space colonies, with their maneuverability and mobility, offer the most efficacious methodologies for *domesticating* solar radiation to provide the basic energy and light requirements of a human colony in space.

The upshot of all this has been the development of an ethos of awareness of the radiation environment in every place within the Solar System.

In 2025 CE, the space colonists began to develop human outposts within the L1 Lagrange Zone of the Sun; two years later, they completed construction of a fully-operation logistical center and space harbor facilities as part of an overall infrastructure, which is designed to support the development of new human colonies on the celestial bodies of the Martian Subregion. In the process, they became more familiar with the challenges of long-distance travel, and of long-count residence, within new regimes of charged particles and new vagaries of planetary weather.

The Martian colonists soon realized that they were entering a new unknown as they moved farther away from their home planet and its territories in "near-space;" a feeling that was not unlike the apprehension that was felt by their Pleistocene ancestors almost two million years earlier when they ventured out of East Africa. But, like their antecedents in the Tropic of Cancer Region, the modern space colonists would find that the Martian Subregion was not significantly different than the Cislunar Subregion (CS), to which they become accustomed.

However, decades later, the space colonists would encounter a very different set of environmental challenges as they sought to colonize the Jovian Subregion, and the dwarf planets of the Kuiper Subregion. There they would encounter an unexpected variety of micro-climates on the moons of the so-called Gas Giants, and on the Dwarf Planets.

But, in general, the space colonists would continue to perceive space weather in terms of radiation from the stars and the Sun, as well as the decay of atoms; and on celestial bodies, they would experience the familiar elements of planetary weather, such as temperature, winds, dust storms, as well as precipitation of exotic volatiles, such as methane.

One way to describe the kinds of weather which humans sometimes encounter on other extraterrestrial celestial bodies, is by comparing it to analog weather regimes on Earth. Thus, the Atacama Desert of South America – where virtually no precipitation occurs – is routinely used to prepare humans for living and working on Mars. Another dry place, the Sahara Desert is used as a model for the global-scale sandstorms on the Red Planet. The cold and dry Gobi Desert is like that which is found on the Earth's moon. And, the volcanic landscapes of Hawaii can prepare humans for planet Mercury, which is pot-marked by volcanoes over its entire surface. Even the water geysers on Iceland have been studied to gain greater understanding of methane geysers on some of the celestial bodies of the outer Solar System.

The "days" and "years" pass on other planets and some moons, just as they do on Earth, but they vary in length. Some planets rotate either slower or faster than the 24-hour period that humans are used to on the home planet. This means that some planets and moons will present the various "faces" of their sphere to the Sun, at various times and for either shorter or longer periods. Also, planets and moons will journey around the Sun in orbits of varying distances and duration. And, some planets and moons will present themselves to the Sun in an erratic way; presenting their faces at various times, depending on their orbital presence among other larger or nearer planets and moons. All these manifestations of celestial mechanics have a great deal to do with the weather and climate that is presented to would be earthling space colonists.

In practical terms, this means that (given the current state of technology), humans cannot survive in the open on any planet except Earth, without the enhancement of technology; they must create small-scale replicates of the home planet on permanent space colonies; that is, extraterrestrial human colonies – whether within the interplanetary medium or on the surface of a celestial body – must be *terraformed*.

Early Warning and Protection

The Geospace Space Weather System (GSWS) has been established to monitor space weather. It does this by tracking solar emissions, both the routine radiation of charged particles, and the occasional eruptions and storms. Magnetic fields and other electromagnetic phenomena also are routinely monitored and reported by a network of orbiting stations.

Depending on the force with which particles from solar eruptions are ejected, it takes between 1 to 8 days for these particles to propagate throughout the heliosphere. This means that forecasters have sufficient time for issuing warnings to space colonists, so they can take measures to deal with harmful effects of cosmic and solar storms. Although it is difficult to predict when a solar storm will erupt; once it does, the movement of the charged-particle field at various points of propagation can tracked. Another phase of the solar wind regime occurs when it confronts and mixes with magnetic fields and belts of charged particles around celestial bodies.

Meanwhile, routine space-weather reporting provides information about the activities of the solar wind and about charged-particle precipitations. Armed with this information, space colonies can use shielding and other countermeasures to deal with the deleterious effects of charged-particle *weather*.

But there are still places within the Solar System where earth-like weather can be found. So, there is rain, snow and fog – which are perceived as either a blessing or a curse, depending on their effects on human activities. Similarly, regimes of heat and cold can cause several degrees of discomfort, or even harm to the human organism. And, winds can be perceived as useful or an obstacle, depending on the nature of human activities that are involved: they can propel spacecraft or damage colonies on the surface of celestial bodies.

Temperature regimes are an important aspect of weather in space or on celestial bodies. This is because humans have been designed for living and working within the parameters of heat and cold extremes on Earth; where temperatures generally occur within the ranges of the comfort zone of the naked human – or a human with a coat, shelter and some sort of artificial air conditioning technological support. And, aside from the effects of extreme temperatures on both carbon-based and silicon-based humans, there is also the ways in which extreme heat and cold affect the state of being of chemical elements in outer space, and on celestial bodies. Thus, as humans have ventured into the higher latitudes or higher altitudes of the atmosphere on Earth (or other bodies), or into space, that the effects of cold temperatures have required some form of artificial protective capsule or cocoon for humans and their operating systems.

Personal Shielding

On Earth, individual humans can protect themselves from the effects of extreme cold by adding layers of *artificial skins* (i.e., clothing) to keep from losing body heat – even in the higher latitudes and altitudes; but in outer space and on extraterrestrial celestial bodies considerably more applications of technology are required, because of the added factor of charged particle radiations and fields.

It is manifestly obvious that some sort of and degree of personal shielding measure is "vital" in outer space and on celestial bodies, especially those which either have no effective magnetosphere or a weak one. The need for personal shielding occurs along a spectrum, which varies from an absolute requirement in the near vacuum of outer space, where every kind of cosmic and solar radiation, at various phases of the electromagnetic spectrum travels in waves, at nearly the speed of light, and with various degrees of penetrability; that is, the ability to penetrate human skin, bones and organs; with a destructive capability to damage or kill cells, either immediately or over a longer duration.

So, as the period of human colonization of the Solar System has progressed, interplanetary space has become increasingly familiar. The early perception of this medium as being a *dark vacuum* has been changing continuously.

Community Shielding

While personal shielding is vital in space colonization, it has become obvious that protection from the effects of *space weather* is more effectively done at the community level, where comprehensive technological systems of terraforming can serve the entire group. It is an approach which recognizes the concept of synergism – the product of community shielding is greater the sum of individual shielding efforts.

This approach has been put into practice on orbiting space colonies by the ingenious utilization (and multiplication) of the constant movement of the colonists themselves within the colony. So, literally, each step and other movement by colonists, as they carry out their daily activities within the superstructure of the colony, is absorbed by the inner bulkhead and floor as kinetic energy. This energy is converted into electrical energy, which also generates a magnetic field on the "ceiling" of the space colony.

Routine community shielding techniques are efficacious in providing protection to space colonies and bases against the harmful effects of radiation from normal cosmic and solar emissions, but to protect against such radiation "storms," more active and energetic measures are required. These usually involve "escape and evade" tactics by orbiting colonies to avoid the most energetic pulses and plasmas.

There are also other techniques being employed within space colonies that take advantage of the synergies that are derived from the multiplication of minute individual forces – such as the barely perceptible individual twisting of knobs – that produce an enhanced combination of kinetic energies – which can be captured, enhanced, and converted into a useful type of energy for doing desired work – in this case, to generate a magnetic field.

Artificial shielding technologies for human communities first emerged as a protective strategy as early as the 20th century as Holocene, as *climate-controlled cocoons* megastructures made of steel, glass and insulation materials to were constructed to provide shielding and climate-controlled environments for major populations. The most significant of these – as far as space colonization is concerned – are the self-sufficient central places that are deployed in inhospitable places, such as the deserts, and the Arctic and the Antarctic.

Also, by the start of the age of space colonization, humans on Earth had already been living and working in large-scale enclosed *colonies,* which were in places where only extremophiles had lived before. They were deployed within the bowels of the caverns, along the intermontane valleys of the highest mountains ranges, and even in the depths of the oceans – as well as in the polar and desert regions. These were special niches, which variously lacked adequate oxygen, or water and food; where gravity negatively affected human respiration, blood circulation and, therefore, normal functioning of the brain, lungs and heart; and where temperatures were so extreme that they could kill a human within a matter of minutes or hours.

These technologies of shielding and *terraforming* have been valuable for colonizing outer space. In this new milieu, however, a different set of environmental challenges are present: instead of water precipitation, there are charged particles that flow through the interplanetary space. These fluxes also create all sorts of space weather field phenomena, as they encounter magnetic fields and other plasmas on their journey through the heliopause. Often these encounters serve to enhance the energy of the particles and to increase their speed, thus also enhancing their ability to penetrate materials and the human cells.

The upshot of the matter is that the electromagnetic and radioactive elements of space weather always and continuously require protective measures if humans are to survive and thrive outside Earth. The human organism simply cannot exist in the space weather environment, where there is no such thing as quality of air. By the same token, on other celestial bodies: precipitation, temperatures and winds almost always occur within parameters that require "air-conditioning." And, there is always the reality of charged particles, which will kill the unprotected human organism – sooner or later. And, without some form of shielding or constant repair, these elements of space weather will either quickly, or eventually, destroy or degrade the operational life of man-made machines and artifacts as well.

Many of the early protective measures implemented by the human colonists in space are *passive*. That is, they are based on the use of *dumb materials* that simply absorb, deflect or disperse cosmic and solar rays. Among these are natural materials such as water, soil, concrete and wood. Artificial materials, which are created by nanotechnologies in micro-gravity conditions also being used in a passive mode. Active protective measures, which involve the use of mirrors and electrical-optic technologies, are rapidly replacing shielding techniques that use natural materials. In any case, the cumulative knowledge of optics and electro-optics is now being used to enhance light and to either focus it, to maximize its effect, or to scatter or deflect it as needed.

Smart materials also are being used to create an artificial outer skin for all manner of human habitats. These include: satellites, space stations, educational, scientific and military bases… and even the largest space colonies. Their function is to detect and analyze radiation threats and (in real-time), and to develop and deploy active countermeasures; with the overall objective of maintaining the structural integrity and operational viability an artificial human system in space.

Active Countermeasures

Space colonists also have developed *active* countermeasures against the most destructive effects of precipitation and storms of charged particles. The most effective of these involves the construction of artificial magnetic fields to surround the space colony, by essentially replicating the development of the natural magnetospheres of some planets, such as Mercury, which although they do not have an internal generating liquid metal core, are able to generate a magnetic field by electrical activity in the subsurface of the body.

Another technique that is complement the magnetic field that is created by the generation of electrical fields, by utilizing the fact that solar-wind plasma does not flow like a normal fluid; it does not envelop the space colony. Instead, as it "splashes" upon the surface of the spacecraft, it develops considerable turbulence, which causes the overall plasma to breakup into smaller "droplets," each of which can be neutralized by being ionized by flows of artificially-generated electrical pulses. Thus, it is now feasible to create a protective magnetic bubble; large enough to protect spacecraft of any size, even one as large as a space colony.

> THE FUNDAMENTALS OF SURFACE-BASED
> COLONIES ON PLANETS AND MOONS ARE
> THE SAME AS OF COLONIES IN ORBIT…

Chapter 25

Planetary Colonies

Personal Systems

Breathing... inhale oxygen for the brain, muscles, and the internal organs... exhale to expel carbon dioxide...

Nutrition... Imbue the brain and body with adequate caloric fuel to maintain optimum consciousness and cognitive functions, and to maintain the anatomical and physiological systems of the body in optimum condition...

Vitality... maintain the optimum electro-chemical interaction between the brain and the rest of the human body...

Balance... maintain the optimum interactive linkages between the brain, the internal gyroscope, and the skeletal system...

Temperature... maintain optimum temperatures within the brain and the body... perspire to cool down all systems as needed... use covering, shelter and heating systems to maintain optimum core temperature of all systems and processes...

Security... create security systems for *personal space*... monitor external environment and internal systems and processes... develop real-time systems to respond to *irregularities* in the environments... use actuators of various configurations to take remedial actions as necessary...

Community Systems

In the short-term, the survival and propagation of a community of humans on a celestial body calls for natural resource extraction and initial processing of raw materials from "captured" asteroids or comets, on-site. In the long-term, however, the plan is to continue using asteroids and comets as sources of raw materials, but to also create the infrastructure for extracting raw materials from the planets and their moons. The same infrastructure will be used to establish proto-type, subsurface space colonies on Earth's moon and on the celestial bodies of the Martian Subregion and of the Jovian Subregion.

The SB humans can do the *extravehicular* work of constructing factories... these manufacturing and assembly units are now doing the basic work of *terraforming* introducing carbon-dioxide and other green-house gases into the thin atmosphere of Mars, with the intention of eventually developing an Earth-like atmosphere and a warmer planet. This process begins with operations to capture asteroids and to use them as resource extraction sites initially, and eventually, as processing and manufacturing centers. The asteroids also can be used as logistical and maintenance depots, as platforms for observatories, orbiting military bases, and so on.

The long-term plan for colonizing other planets and moons, involves such things as melting of polar and subsurface ice deposits. The most commonly discussed tactical approach is to melt the ice through artificial warming of the surface, and thereby, release carbon dioxide into a developing atmosphere. At the same time, the melted ice water would be used to irrigate plant life, which generates additional carbon dioxide... and so on. It is hoped that, through such human technological interventions: water, oxygen, hydrogen, and carbon dioxide cycles might be activated over a long-count interval of time.

But without a global magnetosphere to protect the planet from the dissipating effects of the relentless solar wind, the colonists realize that all the efforts described above will be for naught; unless an artificial magnetosphere and can be activated by embedding electrical charge networks over large swaths of the celestial body. And indeed, there is a plan for terraforming Mars that calls for laying cables just beneath the surface, to create a series of electrical circuits resembling the underwater transoceanic telegraphy cables that were strung out on the home planet in the 19th century – but these cables transmit electricity at various parallels on the surface of the planet, to generate a global magnetic field. The energy for creating the needed flow of electricity is provided by a network of solar panels that are placed strategically in orbit around the planet.

Walter Gomez

TERRAFORMING MARS WILL TAKE A "LONG TIME"

Consider that: the distance from the Sun to the heliopause is about 30 Astronomical Units (AU) and each AU measures about 91 million earth-miles; that the current cruising speed of spacecraft is approximately 20,000 miles per hour; the time it takes an "empty" spacecraft to transit the Solar System – from Earth to Pluto – is on the order of 18 months; the time it takes a Spacecraft-Payload Unit (SPU) to travel from Earth to Pluto (1 AU) depends on the mass-load of the SPU, and the net force of propulsion; the time it takes an SPU to travel from Earth to the other planets depends on more variables, including additional elements of infrastructure at the point of departure, along the mission trajectory, and at the point of destination; and that the time it takes to construct a space base, a space colony, or any structure – in space or on celestial base – depends on a greater number of variables that are involved. Thus, as the scale and complexities of the project increases, the longer the time it takes to deploy a fully-functioning space colony to become operational.

The controlling time variable will be the time it takes to complete the long-count projects, such as terraforming of planets and moons, which can only be measured in terms of 100,000 *Decas* (periods of time lasting 100,000 years); as opposed to less-ambitious projects on planets, which can be measured in terms of earth-time (days, weeks, months and years).

Chapter 26

Diverse Populations

The concept of "extra-vehicular activities" itself is a tacit acceptance of the fact that traditional carbon-based humans are not designed for surviving the environments of outer space or any of the other celestial bodies of the Solar System – without considerable inputs of technology.

Temperatures on Mars, for instance, are generally too extreme for long-term outside activities by carbon-based humans; but never too extreme for silicon-based humans (genetically-enhanced cyborgs or robots). On the Red Planet, temperatures (Fahrenheit) swing by as much as 60 degrees, twice-a-day. In the mid-latitudes, surface temperatures range from 0 to -180; and surface temperatures at the polar caps drop to -200 during the winter, while the warmest soils of the equatorial zone occasionally warm up to +80.

Temperature regimes on Mars can be characterized as being only marginally more extreme than the most extreme conditions on Earth. Indeed, there are places on earth that are as hot or cold as parts of Mars. Consider the Gobi Desert and the lands of the Arctic and Subarctic polar regions; they are as cold as can be found on the Red Planet.

What makes outside activities even less hospitable on Mars for carbon-based humans, however, is the lack of breathable oxygen, the unabated solar winds and the massive and enduring sandstorms. In these circumstances, it has been found that cyborgs and robots are the more suitable *human resource* for carrying out the tasks related to the construction and maintenance of an infrastructure for terraforming other planets. This has been the lesson learned from practical experience with the robotic landers and rovers that have been operating on Mars and the moons of the Jovian Subregion since the latter decades of the 21st century CE.

However, even robots must be protected from the effects of strikes by asteroids and comets, or even meteoroid storms on Mars. At the other end of the debris-mass spectrum, there are the Martian mega-sandstorms that scrape the surface and obscure the atmosphere for months and even years at a time. In any case, all these particulate precipitations have a negative effect on the optimal operation of both *smart machines* (robots) and so-called *dumb machines*. The problem of fouling of movable parts and seals, for example, has been a well-understood environmental hazard for machines in the desert environments on Earth.

The behaviors of the Sun and the occasional activities of the galactic stars will continue to affect human activities in outer space and on a few celestial bodies, to some degree or other, both in the short-term and over longer accumulations of time. Ordinary cosmic or solar events also can cause severe damage to humans, to their structures, machines, and the electric systems that provide power and connectivity to them; in the extreme, they can cause electrical circuits and electric power grids to experience disturbances and even total breakdowns.

Given all the above, it has become almost a *given*, that the first iteration of colonization of the many planets, will have to be accomplished by silicon-based humans. Meanwhile, for instance, as Mars continues to undergo terraforming, other, protected micro-colonies are also being deployed on the Jovian moons. According to an overall plan, SB humans will also construct the first permanent settlements within the Jovian and Plutonian domains. And, over the course of centuries, or even millennia, various micro-environments will be terraformed to such an extent that bio-engineered cyborgs can join the effort to colonize these celestial bodies.

Making the Robots and Cyborgs Smarter

By the turn of the 21st century, the Holocenes had developed the material *hardware* and the *software* to enable robots to do anything carbon-based humans can do in space operations. This technological achievement has proven to be the key to achieving the cosmic imperative of colonizing the whole of the Solar System. The fundamental reason for this is that the natural environments that lie outside the Earth's atmosphere and protective magnetic shields have proven to be inhospitable to carbon-based humans and, therefore, only silicon-based entities can colonize the most distant and, therefore, isolated places – those that cannot be supported by the near-earth colonies.

Consider that, during the two-million years following the Great Human Awakening Event, carbon-based humans have amassed a vast archive of cultural experience in using non-human entities to do the things that were beyond their natural capabilities. As examples: dogs have been used to enhance the tracking abilities of humans in the hunt for prey; the power and speed of horses has enhanced he innate capabilities of humans for moving about the surface of the planet; and falcons and other birds had taken to flight as auxiliary hunters and as carriers of information.

It was only during the last 8,000 years, however, that carbon-based humans turned to non-organic machines to do the things that were too dangerous or beyond their innate strength and endurance.

What made this possible was the discovery of the many secrets and nuances of *stored energy*, and the means of harnessing it to do purposeful human work. Thus: coiled metals and ropes could be released in an instant or gradually, to release stored torque energy; the force of falling water or sand could would release gravity force held in abeyance; and boiling water would release energy in the form of pressurized vapor. And it was found that the atom's nucleus and electrons produced energies as they were transformed from one state to another… by some catalyzing agent.

Homo Spaciens

Consider the case of the newly-developed *Homo spaciens*. These are Homo sapiens that have been genetically engineered to survive and thrive under some of the most extreme environmental conditions of the Solar System. These are a new form of human, whose enhanced skins, organs and bones can fight off the effects of the most dangerous radiation, including gamma rays. They also can self-heal to repair damage caused by trauma of any kind; and their organs and other aspects of their physiology have undergone genetic re-engineering, so they can ward off or repair the damaging effects of living in space.

In the same vein of modifying humans for life in space, the reproduction process aboard space colonies has undergone radical changes as well. In this case, the ultimate objective is to artificially engineer every aspect of the natural reproduction cycle: from the production of eggs; to fertilization; to implantation of DNA; to incubation in vitro and gestation… all of which takes place *ex humana*, in artificial laboratories and incubators. Ultimately, the genetic architecture of Homo spaciens is designed to produce the optimum carbon-based human organism for space colonization.

Nevertheless, these genetically-modified humans are still be incapable of surviving and thriving in extraterrestrial space in *au natural*. They are still carbon-based organisms that require oxygen, ambient pressure, food and water… as well as all the other elements that support the human physiology. None of these are available in the natural environment of outer space.

Still, Homo spaciens are "extremophiles" that can operate effectively in the environmental extremes on the celestial bodies and, to a more limited degree, in outer space. Indeed, that is why they are being used as first-responders to emergencies, within a spacecraft or in extravehicular situations.

Form Follows Function

There are many varieties of silicon-based entities, known as robots or bots. They are designed to perform specific tasks, and their form reflects it.

"Smart" robots work more autonomously than the so-called "dumb" bots, because they do the more intricate tasks and procedures; in situations where preprogramming or real-time control by humans is not feasible.

At some point, "smart" robots become silicon-based humans. The transformation occurs when "smart robots" take on all the properties of carbon-based humans. The process begins when smart robots are made in the image of their carbon-based human creators, so they can operate in the same ergonomically-designed workstations. As a result, these smart robots begin to look and act like carbon-based humans; they perform tasks in the same ways that CB humans do; and they also *learn*... in response to unanticipated cues from their environment and job situation, without any further input from their CB creators; instead of writing copious lines of program code to enable the silicon-base robot to cope with as many of *expected situations* as possible, their carbon-based human creators have taught them by flooding them with mega data imagery of the environments and situations they will encounter – thus imbuing them with a level of prescience which normally only the human mind has. Beyond that, SB robots now actually carry out inquiries and experiments within the framework of formal models, theories, intuitive conjecture, and specific hypotheses. And, *enlightened* robots are adding to the common base of knowledge and skills in support of space colonization.

Thus, one of the main reasons why silicon-based humans have been so successful in operating as artificial surrogates for humans, is the breadth and depth of *foreknowledge* that is given to them during the design and programming phase of their development. Big Data, High Resolution Imagery and Deep Learning techniques and technologies provide them with realistic *a priori* models of the environments and situations that they will encounter in space or on a given celestial body.

In a sense, the SB humans have been provided with the sentience, cognition, logic and intuition that could only be achieved through decades of intuitive insight and empirical observation by CB humans. Thus, for example, SB scientists and engineers are already engaged in the task of building the infrastructure for space colonization, using a combination of computer-generated and virtual reality imagery, as well as the store digitized memories of humanity that is stored in the *Cosmic Cloud*.

The upshot is that the teams of SB humans that are sent to construct the space colonies are already intimately familiar with the environment in which they would be working – not by copious lines of coded instructions or a grand complex of algorithms – but, rather, as the result of a veritable flood of images of all sorts. These are the manifestations of the old saying that "a picture is worth a thousand words".

And, indeed, these SB humans have been prepared for life in space and for the job of constructing space colonies and all their supporting systems. As a result, they can construct the infrastructure of subsurface work and residential spaces, at no risk to CB humans, or the need to expend exorbitant resources just to maintain CB human life during the construction of these habitats.

There is still a complex division of labor in the construction of space colonies. Silicon-based humans work with specialized bots and drones in carrying out normal operations and maintenance functions. And, these auxiliary entities often take on exotic forms, such as the spider-like bots that maintain and renovate the inner and outer "skin" of the space colony. These diverse systems of beings, of entities virtually *swarm* the space colonies with a greater complexity of intelligence, structure, materials, powers, talents, skills and other qualities... all acting interdependently towards the cosmic imperative of colonizing the Solar System.

As the human condition in space has become better understood, it also now obvious to planners and decision-makers that... if the *human* colonization of the solar system was to be achieved... the very meaning of what it means to be *human* would have to be broadened. Consequently, the space societies are now comprised, not only of carbon-based humans, but also genetically-enhanced beings and an assortment of artificial beings, including the part-human-part-silicon cyborgs, and a variety of other sentient, *intelligent* entities...

Simply stated, the natural environments of interplanetary space, and all celestial bodies within it, are either acutely or chronically lethal for carbon-based humans. It is a quasi-vacuum within which molecules of oxygen are so dispersed and whose ambient pressure cannot hold the human organism together. It is also a three-dimensional desert, where no appreciable water molecules exist.

The temperatures in space can be many times colder than the Antarctic when there is no solar radiation, and they become much hotter than Death Valley when the Sun illuminates anything except a body with zero albedo.

Hence, when the Sun shines an object in space, such as a spacecraft during daylight hours, it becomes too hot for the human skin to bear; but when its rays are absent, it is too cold to touch.

Other perils to human operations in space are less obvious. Thus, even within a terraformed cocoon, the effects of prolonged weightlessness on the human body, can be slow and subtle – yet deleterious to the CB human organism less dangerous to CB humans. What this means in practice, is that without some form of exercise and technological intervention, the CB human will experience bone loss and muscle atrophy, especially of the lower portion of the body. And, this wasting of the bone mass also will cause rises in calcium levels in the blood, and consequent incidence of kidney stones. In a sense, the effects of gravity in space the long-term aging process that occurs to CB humans on Earth.

Radiation – which also causes problems for the CB organism – is a constant, and sometimes even an acute problem in space. In the short-term, passive and active shielding systems are adequate protectors of the natural human organism, but as the period of exposure to radiation lengthens, there will be a cumulative and, ultimately, an incapacitating or even lethal effect.

So, to deal with this problem in a definitive way, carbon-based humans are turning to changes in the human organism itself, so that they can be better equipped to deal with the environmental challenges of living permanently in outer space. In practice, such adaptations have included changes in diet, fortifying vitamins and minerals, as well as the use of chemicals to change the biochemistry of the organism, artificially-engineered organ transplants and, the most profound application: genetic engineering.

Another approach to the problem of permanent living in outer space is to create artificial replicates of the human organism. These generally fall under the umbrella rubric of *humanoid robot*, which refers to any artificial, intelligent entity that is designed to perform a task that a human cannot do, either because it is beyond the inherent power of the human or because it is too dangerous for the human organism.

Some robotic systems continue to carry out the space exploration missions. But, they also are being instrumental in the building and maintaining of space colonies and the infrastructure that supports them. So, while some robots (and their bots) workers are assembling space buildings, others are busy creating the infrastructure that will eventually support large populations on space colonies. Put concisely, some robots prepare the way for the migrating carbon-based human colonists, while other robot workers will continue to function in auxiliary ways throughout the life of the space colony.

Then there are those that become silicon-based humans…

Silicon-Based Humans

In some cases, robots take on a human form – in the fullest sense: they look like the carbon-based humans; they move the way that is expected for CB humans, and they even emote in the expected way that CB humans would in any given situation. In other words, they "fit in" nicely in any situation aboard a space colony. But they are different in one important way: they can withstand more physical and emotional stress than carbon-based humans. This makes them indispensable in times of emergency situations that involve physical or psychological stress.

These *human-like* robots are about the same size and weight of the average carbon-based human colonist, by design. The main reason for this is so that the robot can perform its assigned tasks in situations where the work environment has been basically designed for human operatives. And, another advantage of having a robot with a humanoid form in space settings is that it can be effectively integrated onboard spacecraft; their size and other aspects of form means that SB humans can conduct any of the various tasks that are normally assigned to CB humans, in all the and spaces of the colony that are normally designed for to their creators.

Humanoid robots have soft skins and expressive eyes and mouths, and all the other outer aspects of humans; there is no cold, hard metal or plastic epidermis to set them apart as being a machine; it feels "natural" to be around them and to interact with them. They feel like humans, behave appropriately as humans, and express expected emotions that are appropriate to a given situation.

Endowed as they are: SB humans continue to help humans explore and colonize the solar system in every other way; they do the work of exploration and scientific studies on planets, moons, asteroids and comets, as well as the interplanetary medium, and the processes and systems that occur throughout the solar system, and especially in places that are too dangerous or distant for humans.

But their ability to live and work in any natural environment of the Solar System is derived from more than just their humanoid anatomy and physiology; they also have developed a human-like brain and even a holographic entity which operates like the human mind. This makes them invaluable as "scouts" that can reconnoiter conditions on a planet or other celestial body of the Solar System that have not yet been surveyed by landers or rovers. In a sense, this is akin to the intuitive precognition of the human mind that can imagine what future conditions will be in the unknown. These qualities have enabled planners to design exploration and colonization missions more quickly and with greater confidence of success than by the traditional methods of orbital mapping and the use of landers and rovers.

Sentience and Intelligence

Most robots are not "humanoid." In fact, the first robots in space were *smart rockets*, which were essentially converted military Intercontinental Ballistic Missiles. These long-range ballistic missiles are considered "smarter" than the original *dumb* rockets because human intelligence is imparted to them – in the form of pre-written or real-time instructions – so they can operate, with a variable degree of autonomy in making *enroute* changes in direction, velocity and other navigational parameters.

Now, in the age of space colonization, another class of *smarter* robots has emerged to operate throughout the solar system. These are sentient and intelligent robotic emissaries that are being sent into the interplanetary medium and sometimes deposited on the surface of the planets and other celestial bodies.

This class of robots has become quite diverse in form and function. Sometimes they are called satellites; others are simply referred to as probes, orbiters or landers or rovers; some perform their functions *off-shore*, as it were, as they orbit the Sun or some other celestial body, while others do their assigned missions in the atmospheres, surfaces, and subsurfaces of bodies.

Within the class of orbiters, there are those that operate as "servers" within the networks of communication; or as beacons (e.g., GPS) for navigation and as reference points for determining locations on the surface of planets and other larger celestial bodies; and others are serve as air reconnaissance systems. In any case, what they all have in common is that they receive instructions – either directly from other humans or via computer devices.

SB explorer-scientists also are being sent out: to explore and survey the heliosphere and the developing Geospace Regions to study the Sun: its structure and processes; its core and the intermediate layers of magma and churning and fluxes of charged particles…

They study magnetic fields and plasmas; the structure, composition and processes of all other celestial bodies, which range in size from the meteoroid, the comet, the asteroid… to the planets and their constellations of moons…

Others are focusing on specific planets and moons as candidates for human colonization…

And, swarms of SB explorer-scientists also can be found within the Main Asteroid Field and the Trojan Zones; striving to develop comprehensive data fields, with the objective of integrating these resource-points with processing, manufacturing and other specialized nodes within the Geospace Region.

Silicon-based robots and bots take on many specialized missions and tasks, and their form reflects this diversity. The more complex the mission, the greater the sophistication of form and processes. As a result, the humanoid or arachnid SBs which operate on space colonies and space bases are not only larger; they are also more complicated. They carry within them integrated systems for extracting geological and other samples from the atmospheres and surfaces of celestial bodies, onboard laboratories and data processing centers, and communication functions

A major advantage of using SB rovers on celestial bodies lies in their ability to cover large areas of space in relatively short periods of time; to reconnoiter and study; to process data in real-time; and to transfer information to any user throughout the Solar System, all in real-time. By contrast, consider the centuries it took the early Holocene colonists to gather sufficient data points from the Earth's surface, atmosphere, and waters to gain the comprehensive understanding of the planet; compare that with the vast quantities of data that SB rovers have already acquired and processed in less than one century since the first humans landed on the Moon.

Degrees of Autonomy

The questions regarding the nature and degree of control by humans should maintain over robotic entities in space operations have come into greater immediacy and focus as the robots have begun to transcend their original work as remote data gatherers. Originally, they almost always were under the direct control of a human or the indirect control of a human programmer. Thus, the earliest space probes and artificial satellites operated in accordance with hard-wired instructions or pre-written software coding. Gradually, however, the robotic entities were imbued with software that included discretion at certain Boolean decision points which are presented during the course of a long and complex mission.

Today's robots are working as astronomers, geologists, physicists, and engineers are being assigned to work on the new colonies along the frontier of the Solar System. Generally, their missions are so complex and so filled with uncertainties, that the robots must be equipped with the same degree of intelligence as carbon-based humans. Consequently, they are being routinely imbued with ever-increasing powers of learning and logic, especially when their assigned missions involve many more decision points than ever – whose solution depends on powerful precognition and intuition.

Consequently, it is now a generally accepted fact, that the ability of the space colonies of the near-earth subregion of the Solar System – to maintain effective lines of communication and to provide logistical support to the new colonies – has become so tenuous that greater autonomy must be provided to the silicon-based colonists of the outer regions. In practical terms, this means that they must be imbued with the same degree of intelligence and prescience as the carbon-based humans – thus, they have become "humans" in every sense...

Such an evolution of circumstances continues to beg question of how much autonomy should be ceded to the robots. Consider that the initial robotic conveyors of *payloads* into space faced relatively few decision points in accomplishing their assigned missions. Relatively linear and comprehensive algorithms, including guidance and control commands, could either be generated by a ground-based computer or by an onboard computer. And, if unexpected events occurred during the mission, a human controller could intervene with *managerial overrides*. But now, as the number of space robot missions has multiplied, the complexity of the missions has increased by magnitudes of order, the time and distance parameters of command and control have expanded and, therefore, some form of *human-like intelligence* presence is required *in situ or on-scene.*

And, since it has become acutely obvious that *carbon-based* humans are limited by physiology and psychology for long-term residence in space, the concept of what it is to be a *human* space colonist has now greatly expanded; robots have become *silicon-based humans*, who are now wholly integrated within the *society* of the space colonies. They are *human-like* in both appearance and behavior… and they are programmed to experience the same feelings and emotions as carbon-based humans. Thus, SBs are now considered as just another element of diversity in the population of the space colony.

This reality of acceptance of the diversity of human resources has now become a powerful factor in the ability of earthlings to colonize every subregion and planetary environmental niches where carbon-based humans are not able to function effectively.

Chapter 27

A Low-Mass Strategy

The strategy that has been selected for colonizing the solar system can be described as a "low-mass" one. Thus, instead of establishing colonies on the surface of planets and moons, or even on asteroids – which are "high-mass" bodies – it has been decided to deploy the first generation of space colonies within the interplanetary medium, where the gravitational field is relatively smooth, unlike the more turbulent conditions of the planetary depressions.

Ultimately, by locating space colonies *offshore* – beyond the influence of the gravitational wells of the more massive celestial bodies – space colonists can realize the benefits that derive from dispersal and mobility as a means for ensuring the ultimate survival of the human species throughout the universe.

The low-mass strategy for deploying human colonies throughout the interplanetary medium has several advantages over a high-mass planetary one: (1) it minimizes the energy requirements and the associated costs of launching and landing spacecraft; (2) it optimizes the ability of space colonies to avert impacts by meteors, comets and asteroids, because of the greater adroitness of movement in outer space; (3) it enables space colonies to rotate and move in various directions in order to maximize the benefits of sunlight, and to minimize the negative effects of cosmic and solar radiation; and (4) it provides space colonies with manageable gravitational *throttling* capabilities for gaining access to raw materials on asteroids and comets.

One logical corollary of the low-mass strategy might be that there is no need to establish permanent human settlements on the planets and moons; which are more massive, and therefore have deeper gravity wells. (The deeper the gravity well, the more fuel is needed to reach escape velocity). However, less massive celestial bodies still can be utilized as resource sites, because the value of the resources they contain exceeds the higher costs of extraction and transportation that is involved.

Another advantage of the *low-mass* colonial strategy is that that transportation within a network of low-mass nodes is less costly and easier to manage than a network which contains higher-mass nodes. Aside from the relatively low-cost of launching and recovering spacecraft at each cosmodrome, the relatively low gravitational *displacement* of the artificial orbiting space colonies makes it easier to navigate the forces and perturbations within the gravitational *ocean*. And, continuing with the maritime metaphor, the constellations of orbiting space colonies are like artificial *islands* lying off the shores of *planetary landmasses*.

Just about every aspect of the design and planning for space colonization is in keeping with the overall low-mass strategy. The materials used in the manufacturing of the spacecraft – the outer skin and the various structural components -- are made of light but strong materials, with an eye towards withstanding the forces of the space environment and the stresses imposed by them. Also, new *exotic* materials, composed of artificial combinations of molecules that are not found in nature, are custom-made for a specific space environment. Many of these are also *smart* materials: they contain micro-sensors and actuators, which can change the composition of the surface materials in response to changes in temperature and radiation conditions. Other customized materials are fabricated to deal with the extreme temperatures that are created by of bottle rockets to modify orbital courses.

Consider that such advances in materials science are possible because of cumulative advances in the science of chemistry over many generations of Enlightened Minds. The iconic expression of this is the so-called Table of Periodic Elements, which catalogues all the *known* natural and unnatural elements of the physical world: the *building blocks* of matter that can be reduced no further. With these computer-generated combinations of elements, the architects and engineers have been able to create customized spacecraft and space colonies.

The low-mass strategy has also been manifested in the architecture of propulsion systems, which includes the use of materials that are designed and manufactured at the molecular level; these are "smart materials" that can detect and respond to variations in temperature and radiation regimes, as well as in gravitational environments – autonomously and in real-time.

Also, even as space colony structures have been getting larger, their propulsion systems have been diminishing in size and weight. One of the main reasons for this are the continuing increases in the efficiencies of propulsion technologies and methodologies. A basic factor in these greater efficiencies has been the development of small and light, but extremely efficient and powerful battery packs, which harness the energy of solar radiation in normal operations. Acute maneuvers, which require acute surges of power, are energized by nuclear fission or fusion technologies.

The bottom-up, low-mass strategy is also an important consideration in the development of the logistical systems that support the orbiting space colonies. It is a recognition that the greater the cumulative mass of propulsion systems, spacecraft and payloads – the greater the "cost friction" will be in transiting and transporting people and materiel throughout the Geospace Region.

In its practical application, the less dense asteroids and comets serve as important nodes in the networks that serve the space colonies. They function as: (1) as military bases to ensure the common security of the developing Geospace Region; (2) as scientific stations to do the continuing research to increase the efficiency of the orbiting colonies; (3) as observatories for studying and monitoring the movements of natural and artificial bodies; (4) as processing and refinement centers, with the aim of minimizing the cost of providing raw materials to the orbiting space colonies; and (5) as nodes in the transportation and communications network.

This alliteration of functional nodes in the Geospace Region is better understood when they are viewed as being "smart" nodes in an interacting human spatial region. Thus, a cluster of nodes, such as the Cislunar Subregion, is best depicted as a system of interactive and interdependent nodes, each having a contributing function towards the ultimate objective of the system. Thus, for example, the Cislunar Region can be appreciated as a spatial, functional system consisting a constellation of military bases, research and development institutes, universities, observatories, and so forth, that are situated near a constellation of orbiting "bedroom" space colonies.

Dispersal and Reintegration II

Consider the inundation of a presumptive continent where the Malay-Indonesian archipelagos exist today, and the subsidence of Atlantis west of Gibraltar: both were caused by global warming and flooding events that occurred during the Late Pleistocene and Early Holocene periods.

Those events, in turn, caused the dismemberment of the existing human genome into isolated genetic pools. But they also provoked counter-phenomena that resulted in the subsequent remixing of genetic pools, ultimately resulting in a new global human genome on Earth.

The lesson that has been learned by space colonists from the Late Pleistocene existential crisis is the need for dynamic and dispersed pools of human genetic materials that can adapt and respond efficaciously to the various and changeable natural environments that exist throughout the Solar System.

It is also recognized that space colonies are separated by vast distances and that, therefore some human populations will be isolated, and that, therefore, unique and indigenous evolutionary adaptations could emerge within the stranded populations.

The steps that are being taken to diminish the likelihood of extirpation or extinction of human populations in space include a program to enhance the genetic mutation process by artificial means to precipitate evolutionary adaptations to changing and variable natural environments. Thus, advanced genetic engineering and controlled gestation and incubation within controlled laboratory environments, are used to designed to produce a *Homo spaciens* variety of human, one which is naturally equipped to survive and operate in the more extreme *alien* environments of outer space and the celestial bodies.

Meanwhile, as the number of space colonies, constellations and their linkages has grown, the chances of genetic isolation also have diminished. The reason is that the genetic pools aboard individual space colonies and constellations of colonies are easily integrated by transporting low-mass genetic material payloads throughout the Geospace Region. It is a new expression of genetic cross-migration – in the pattern of the intensive reintegration of Late Pleistocene genetic pools during the cataclysmic events of the last period of global warming and flooding.

The Homo spaciens have made traditional earthling concepts of race and ethnicity virtually obsolete. Skin, hair and eye color and facial features are no longer used to create artificial divisions within human populations. This eliminates one of the major elements in creating divisions among populations on space colonies. Indeed, the variety and diversity of form and physiology have become internalized in the psyche of the space colonists.

Cultural differences are also being homogenized by the emerging universal culture of the Geospace Region. Once again, a shared suite of expectations, experiences, beliefs, and a common language are serving as the adhesive of cultural unity. Most importantly, Homo spaciens have internalized the fundamental truth, that the survival of the species on earth and anywhere in the Geospace Region will depend on the shared resolve, that there shall be no more wars and that differences shall be resolved peaceably and through deliberative consensus.

This new generation of humans in space are beyond being globalists; they are cosmologists, and they recognize the proper place of Earth and earthlings in a physical construct which includes the Solar System, the Galaxies and the Universe. Within this context of self-identity, the denizens of the orbiting colonies and those of the planets and other celestial bodies all consider themselves to be of one species... Homo spaciens. This internalization of what it means to be "human" will be vital to the continuing existence of the species throughout the diverse regions of the Solar System, and on the various natural environments on celestial bodies. Seen from this perspective, diversity is vital for the continuing survival and propagation of the human species, now more than ever before on Earth.

Low-Mass Regions in Space

The earliest expression of the low-mass approach to space colonization is the architecture of colonial system in space; which is designed to function as a *spatial system* of interacting nodes, working together to develop the first human political-economic-social *civilization* outside the domain of the home planet of Homo sapiens.

[Each colony is designed to contain diverse human populations of 100,000 CB/SB humans and cyborgs, as well as an adequate population of robotic entities. They all contribute their unique qualities to operate and maintain the infrastructure and the various systems of the colonies].

The space colonies consist of several orders of linkages. These include: inter-colonial networks which connect the colonies within a given constellation; the inter-constellation networks which connect constellations; and the regional networks that connect constellations to the planets and their subsidiary celestial bodies.

This configuration of layered networks provides ongoing interconnection for the existing nodes – at all levels – and provides the flexibility for adding nodes to the spatial systems.

An integrated information system network provides the framework to capture, store, manipulate, analyze, manage and present all types of data and imagery to users at every point in the new Geospace Region. This interactive information system enables users to pose interactive queries, analyze spatial information, and to develop dynamic *real-time* charts and maps, in the form that is most useful to the user.

Now that the basic strategy of low-mass colonization has been generally accepted, the engineers, planners, and policy-makers, are implementing it in furtherance of the overall imperative to colonize the entire Solar System. Thus, they are continuously identifying new places within the interplanetary medium where space colonies can be deployed. And, in doing so, they are continuing to discover new niches that appear to be ideal for development as nodes in the developing Geospace Region. Among the most important of these are the Lagrange Zones [Lagrange Points]. These are areas of relative calm which are created by the dynamic and competing gravitational influences of the celestial bodies in their vicinity. These "harbors" were first discovered during the missions of exploration of the Solar System.

Geospace 2060

AT THE MACRO LEVEL, THE HUMAN GEOSPACE REGION IS BEING INTEGRATED WITH THE NATURAL SOLAR SYSTEM ...

Chapter 28

Orbiting Space Colony Strategy

Establishing a network of space colonies in low-orbit around Earth was the first step in developing the Geospace Region. There are several reasons why this decision was reached, but one of the most important considerations was the overall *cost* of launching rockets and their payloads – which varies according to the mass of the celestial body or orbiting platform from which they are launched. Within this paradigm: asteroids and comets, and the orbiting colonies, are better launch platforms than the planets and moons – orbiting platforms are best.

Fortunately, humans had had several decades of experience in developing orbiting platforms that contain habitats and work stations, such as the International Space Station. Such experiments in long-duration living and working in space, has taught earthlings how to effectively *terraform* artificial spaces within the interplanetary medium. This includes development of systems for providing oxygen, water and all the other physiological and psychological necessities of human life.

Another important lesson that has been learned from the experience with platforms like the International Space Station is how to effectively navigate the ocean of gravity in outer space.

The interplanetary medium of the Solar System; what is commonly referred to as "outer space" can be considered as being analogous to the oceans on Earth. Within this model, the planets and moons, and the asteroids and comets are like islands that are surrounded by outer space.

However, there is a fundamental difference between the islands of the oceans on Earth and the "celestial islands" of the Solar System. That is, unlike the islands of the terrestrial oceans, the celestial islands are always moving. The planets and moons always move along a variety of orbits around the Sun and, simultaneously, the moons are in permanent orbit around a "mother" planet, and comets sometimes take up temporary orbital residence around planets, on their longer journeys from the Kuiper Belt towards their destiny with the Sun. Asteroids sometimes leave the security of the constellation within the Main Asteroid Belt when they are suddenly drawn towards the inner region of the Solar System as the result of jostling by one or more of the planets and larger moons of the Jovian System.

And, there is always the probability of a major shakeup in the movements of the celestial islands as a result of a cataclysmic near miss or actual impact between more massive celestial bodies, such as when a traveling asteroid happens to intersect the orbit of a planet or moon.

So, in the case of the gravitational "ocean" in space, each island is constantly moving and rotating, along an orbital path which is assigned to it by the laws of the field of gravity. Also, each island causes a depression in the gravitational ocean – the depth of which depends on the magnitude of its "displacement" of gravity.

It then follows that the transportation network in space will be constructed as a three-dimensional structure in a celestial *ocean* – whose main nodes will be the cosmodromes, which serve as harbors for the celestial "islands…" from which *space vessels* are launched and recovered. Depending on the depth of the gravity-well of a celestial body, the cosmodrome will be situated on the island itself, or on an outlying structure. This is in recognition underscores the importance of the gravity-well in the *sailing* the ocean of gravity.

[Lagrange Zones originally were known as Lagrange Points, but because they encompass large areas within the interplanetary medium, here they are called "Lagrange Zones"]

Inherent in the overall strategy of low-mass colonization was the decision to develop a network of artificial space colonies where the gravitational field is shallow; where the net attraction of nearby celestial bodies is minimal. One set of such places are the Lagrange Zones. These places occur naturally in spatial situations where two or more massive bodies (e.g., planets or moons) engage in a gravitational *tug of war*, and thereby create an area of dynamic tension, within which the forces of gravity are essentially neutralized by the competing forces of gravitational attraction.

The Lagrange Zones are like natural space "harbors" where the perturbances that are caused by major currents and prevailing winds, in much the same manner as the ocean currents and winds on Earth. The solar wind is dominant in much of the heliosphere, but beyond the realm of the Kuiper Belt of dwarf planets and resident comets, the solar wind begins to interact with the interstellar winds from the Milky Way galaxy, within the outer frontier zone known as the heliopause.

Thus, the Lagrange Zones are places of relative calm in an otherwise turbulent gravitational field. They are therefore excellent places for stationing a fleet of space vessels to ply the space lanes for commerce and to maintain security from threats by nations or terrorist organizations. This highlights the fact that, although there is inter-colony warfare, there is still a continuing threat of attack from earth-based terrorist organizations which possess the minimum capability to launch a missile into low-earth orbit, or which can "hijack" legitimate commercial spaceflights.

Lagrange Clusters (LC)

Lagrange Constellations (LC) are clusters of space colonies which are located at the various Lagrange Zones within the overall Solar System. The relative stability of LCs provides many advantages in observing and monitoring both natural and human activities within their sector of space. They are excellent platforms for monitoring solar activity and the intermittent cosmic activity that affects the space colonies and infrastructure in their domain. And, like the ocean harbors on earth, they serve as important central places and transportation and communication nodes as well.

Lagrange Zones also serve as way-stations within the Interplanetary Highway System, where logistical and maintenance support is provided to spaceships in transit to the outer subregions of the Solar System. And, because of their location relative to the planets and moons, LZs also provide comparatively easy access to a variety of orbits within the constellation network. And, the LZs have become excellent *cache sites* for storing propellants and other essential materiel for longer-distance trips.

Several exploration and reconnaissance satellites are in use LZs as platforms, whose functions include the ongoing reconnaissance of all energetic particle phenomena within the IPM, and especially the solar winds and eruption fluxes that flow within it. The real-time data produced by these satellites is used to monitor the charged-particle environment and to provide warnings of solar storms – not just for Earth, but for the space colonies and other bases as well.

As has happened throughout the thousands of millennia of Pleistocene and Holocene experience, the human mind foresees intuitively what the human brain later perceives empirically. Long before the age of space exploration had begun, there were already Enlightened Minds who foresaw the existence of human space colonies. This precognition is often propagated in the form of *science fiction* words and images. And, sometimes, these precognitions have become reality.

Since the start of space colonization, ordinary humans have begun to *see and experience* the Solar System with greater clarity and resolution – thanks to the artificial sensory enhancements created by the new breed of space scientists and engineers.

Thus, modern humans have come to gain the same intimate level of practical *familiarity* with space technologies that ancestral humans achieved with fire; they now realize, practically, what the human mind has known intuitively; and now the human brain that is actualizing the intuitive realities of the human mind.

For one thing, during the last half of the 20th century and in the first decades of the 21st century, earthling colonists achieved significant practical experience in utilizing Lagrange Zones (LZs) as sites for spacecraft, telescopic observatories, laboratories and communications stations…

An important lesson derived from these early LZ operations is that, as is the case with ocean harbors, the utilization of space harbors requires continual maintenance of, not only to the LZs themselves, but also the orbital position of space colonies and other spacecraft within the harbor. This means, for example, that internal propulsion-throttling systems are used to maintain the equilibrium of space colonies and other spacecraft within the harbor, in response to external perturbances or threats from transiting asteroids or comets.

The Lagrange Zones (LZ)

Overall, there are more than 20 multiple-body situations where the competing gravitational pull of neighboring bodies has created areas of relative gravitational calm; space harbors in which gravitational perturbations are minimal. These so-called LZs occur in various "off-planet" positions, where they are held in dynamic suspension – which results from negotiations between competing attractive forces, within the gravitational neighborhood.

In these neighborhoods, two or more massive celestial bodies dynamically negotiate the location and orbit of five LZs. Beyond that, the entire LaGrange Constellation orbits the Sun in unison. Finally, it should be noted that each LZ really is an extremely large area; rather than a "point". Within the LZ, many of space colonies can be located.

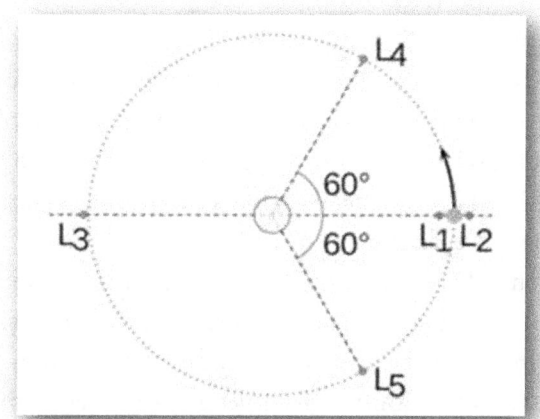

THE TYPICAL LAGRANGE CONSTELLATION

Lagrange Zones Locations

    Sun-Mars   Earth-Moon
    Sun–Jupiter   Saturn-Tethys (Saturn moon)
    Saturn–Dione (Saturn moon)
    Sun–Uranus   Sun-Neptune

Gravitational Order

Every object in the Solar System revolves in disciplined unison around the Sun. This order is primarily imposed by the overall gravitational hegemony of the Sun; and, secondarily, by the slightly lesser influence of Jupiter and the other "Gas Giants". And, on a third-order, regional scale, the planets and their moons also exert local gravitational influence on their less-massive neighbors.

Generally, the more distant from the Sun a body – the greater the gravitational influence on it from nearer, relatively more massive bodies. An example of this is the Jovian Subregion, where Jupiter, Saturn, Uranus, and Neptune – individually and in combination – exert significant gravitational hegemony over smaller bodies that occur within their respective and their combined net sphere of influence. And, in the region of Jupiter and the other Gas Giant planets – which are of more recent planetary construction – there is a population of resident asteroids, meteoroids, rocks, dust and other detritus that adds complexity to the overall gravitational milieu.

Despite the many local variations in the relationship among masses of differing magnitudes, there are some generalizations that have guided humans in exploiting certain *gravitational anomalies* as places to establish colonies in space. An example of this, are the LZs, which are oases of gravitational calm. Many of these are excellent sites for deploying human presences in space. Some are useful in their natural state, while others, such as the so-called L1, L2, and L3 points, are relatively unstable because of their susceptibility to minor perturbing influences, but they can still be managed through a variety of orbit-maintenance techniques, such as rocket throttling.

Some LZ volatilities require more human intervention than others to be useful as harbors; yet they all have integral advantages, such as their: (1) access to solar radiation; (2) access to natural resource sites; (3) access to other LZs, moons, and planets; and (4) access to transportation and communications *highways* in space... to name just a few.

Lagrange Depots

"Accessibility" to water and raw materials is another important consideration in locating both orbiting space colonies and planetary colonies. In the case of orbiting colonies, there are plenty of asteroids and comets that contain these resources, and which are relatively accessible. Lagrange Zones, as it turns out, are also excellent places for developing logistical and transportation depots to the more distant regions of the Solar System, out beyond the Asteroid Belt.

Fortunately, orbiting space colonies now possess the necessary propulsion and navigation systems to maneuver the colony with respect to adjacent natural bodies for resource exploitation. These maneuvering capabilities give the planners great latitude in choosing the sites of colonies. Thus, for example, space colonies can apply the art and science of navigation and maneuvering to position themselves relative to asteroids or comets, to gain optimum access to provide water, minerals, and certain elements or chemicals. Similarly, space colonies can be moved to position themselves near space bases that do manufacturing and other secondary economic activities.

Space colonies also position themselves for optimum access to orbiting solar panel platforms, to ensure the optimum amount and duration of insolation to the colony system. The same dynamic holds true with respect to access to other auxiliaries that are needed to maintain optimum operations on space colonies.

Space colonies must be aware of their *absolute location* and the *relative location* within the Geospace Region. In the case of the Lagrange Constellations, the absolute location variables are centered on the internal gravitational characteristics of the proposed site. The relative location considerations include the nature of the gravitational neighborhood and the relative location of resource and auxiliary nodes within the neighborhood. In some cases, the relative location of a space colony is expressed in terms of the cost of transporting payloads between the relevant nodes of the developing human-based regional system.

Another important criterion in selecting a site is the *gravitational stability* of the LZ itself. The experience of the space colonists has been that the L4 and L5 points are normally more gravitationally stable than the other three Lagrange Zones. That means that space colonies that are in or near the L4 and L5 points require less artificial *station maintenance* to remain in the desired orbital path, than do the other three points. On the other hand, there may be reasons for selecting a less-stable LZ as the site of a colony and its auxiliary bases; and reasons which make it worth the additional orbit-maintenance costs at the L1, L2, and L3 points.

Consider that the Lagrange Zone present the advantages of gravitational stability that the planets and moons offer, but without the disadvantages related to deep gravitational wells inherent in the gravitational tug from the barycenter of more massive celestial bodies. In practical terms, this means that the launching and landing of spacecraft and payloads from Lagrange Zones requires less energy, engineering, infrastructure and, therefore expense than from the "friction" imposed by high-mass celestial bodies.

Chapter 29

Terraforming Space Colonies

The first phase of space colonization began in the year 2018 CE, when the design and construction of the first artificial cities destined for near-earth orbit began. However, even as early as the middle of the 20th century, the knowledge, technologies, skill-sets and the organizational protocols necessary to create self-contained, self-sufficient human *places* in orbit around Earth and other celestial bodies had begun to be developed.

Along a longer trend line, planners began designing structures as precursors to the orbiting space colonies as early as 4,000 years before the present era. Consider the three-dimensional artificial structures that were constructed by early Holocenes (e.g., the Egyptian pyramids); and later, the medieval cathedrals, and modern enclosed urban centers of the 21st century CE. The latter are virtually hermetically-sealed, self-contained capsules in which humans can live, work and play. So, even within the most extreme environments of Siberia and the Arabian Desert, humans now live in artificial *terraformed* spaces which are effectively isolated from the external natural environment.

These are insulated spaces in which the oxygen, carbon-dioxide, and particulate field is carefully monitored by an array of sensors and controlled by electro-mechanical air conditioners, scrubbers and other systems – all for maintaining a *healthy atmosphere* within the artificial space – regardless of the conditions of the external natural environment. Thus, humans in one of these space colonies would feel like they had entered a modern all-purpose residential and shopping mall on Earth.,

So, the main difference between the Earth-based "capsules" and those orbiting within the interplanetary medium, is that the latter require not only artificially controlled temperature, but also a modulated gravitational ambiance, and an artificially produced oxygen/carbon dioxide cycle as well.

Another difference is that space colonies can control the degree and duration of the Sun's light and energy it receives. This proves useful for managing the caloric and nutrition units that are produced by the onboard hydroponics, by controlling the amount of sunlight they receive; either by rotating the entire colony or by using mirrors and "smart" filtering panels to control the quantity and duration of sunlight.

Modular Construction

Since the turn of the 21st century on Earth, large-scale space structures have been almost exclusively constructed by assembling prefabricated modules on-site. The experience gained in the development of materials, components and various operational subsystems, is now being applied to the construction of orbiting space colonies, science labs, observatories, micro-gravity manufacturing units, etc.

Increasingly, orbiting space colonies are designed as special nodes within a constellation of such nodes, along with other nodes that are dedicated to agriculture, processing of raw materials, manufacturing… as well as security and other community support functions. Most, however, are designated as residential and *neighborhood* nodes, which contain all the elements of the modern urban life on earth, including residences, retail shops, restaurants, barber shops and hair salons, and so on; all the places that make urban life appealing to humans on earth.

The iconic precursor of the orbiting space colonies is the International Space Station and all its auxiliary and support systems. Its modular architecture, with separate living and working spaces is the model for all modern orbiting and planetary space colonies within the Geospace Region. It also has been an effective platform for continuing development of technologies and, perhaps more importantly, maneuvers and techniques for loading and off-loading of colonists and astronauts on orbiting space platforms. By the same token, the ISS and other orbiting testing facilities like it, have proven invaluable in the providing practical knowledge and skills required to develop an effective and efficient interaction facilities for transporting large numbers of people and materials between orbiting space colonies, Lagrange Zones, asteroids and comets, and the planets and moons of the developing Geospace Region.

"Operations Normal"

The first order of business for the CB space colonists, upon arriving at a new colony, is to activate all systems for normal operations, and to establish the interconnections with the other colonies of the constellation. By then, the SBs, will already have completed the construction of the structures, and the installation of the complex of internal systems (e.g., navigation protocols, power, communications, life support, *terraforming*…). But the controlling variable for the time needed to achieve ops normal status is the size and complexity of the mission

During this *shakedown phase* of colony operations, virtually all tasks can be subsumed under the rubric of *left-brain work* and *manual work*. Operations are generally controlled by pre-written software instructions to the bots, which do virtually all the extravehicular work on the external shell of the space colony structures; and all the work of maintaining the inner bulkhead and the "ceiling" of the space colonies. Typically, these bots (which resemble earthly spiders) are constant denizens of both the outer and inner sides of the "skin" of the colony structures.

Also during the startup phase of space colony operations, SB and CB systems engineers will be installing the hardware and software that will drive all the life-support and operations systems. The latter will include the early-warning and communications systems that will enable the colony to recognize external threats, such as solar flares and other cosmic eruptions, and to communicate with other space colonies and Earth bases to warn them and to coordinate mutual assistance activities.

This is also the time when the "lifeboat" systems, for both onboard and extravehicular catastrophes, are stress-tested, either by computer-based modeling, or through practical rehearsals, to ensure that residents of the space colony will be able to move to safe quarters aboard the space colony; or move to another space colony in the event of an apocalyptic threat, such as an extremely powerful solar flare or an imminent encounter with a swarm of meteors, or even a hostile attack from terrorist organizations.

Making Colonies Livable

One important factor in making the space colonies viable as self-contained *human places* – whether in orbit or on the surfaces of celestial bodies – is the *sensor and actuator* system. These are crucial for maintaining environmental awareness, especially with respect to cosmic and solar radiation, and early warning of a variety of threats, including from space rocks ranging in size from dust particles to meteors to asteroids and comets. They are integral to the ability of space colonies to, not only scan for threats, but to take corrective actions, automatically and in real-time.

Aside from the need for security, human colonists in space also need the basics of community life. To this end, asteroids and meteors have proven to be useful as sources of oxygen and nitrogen, and water, as well as minerals, although there are also technologies and facilities available for creating within the space colony, as a form of backup. In short, all the needs of a human population in orbiting colonies is met from the "natural environment" (i.e., asteroids, comets, and other nearby planets) or – if necessary – they can be made artificially by using nanotechnologies and chemical technologies.

But one does not live by oxygen or bread alone... The colonists realized early-on that living in an orbiting *city-state* in space requires more than a replicate of earthly material aspects of life to maintain a vigorous and propagating human society in space. There also must be a replicate of Earth-societies: physiological and psychosocial environments that elicit the same *wellness and emotional* state that earthlings feel when they look up into the sky, or feel the power of a redwood forest, or the poetic expanses of desert sands, or an alpine scene, and so on...

The human colonists in space also seem to need many of the *feelings* of shock and awe that awareness of the power of nature inspires to maintain a certain edge – a feeling of animation or anxiety to relieve the monotony that can occur in a space. It is about maintaining an esprit of life within a space colony. It is also the feeling of acute alacrity that is evoked by such earthly phenomena as lightning strikes and tornadoes, or the mundane variability of weather and the cyclic changes. All this is a way of expressing the idea that it is important to make earthlings *feel at home* in the interplanetary medium or on a celestial body.

If terraforming means making earthlings *feel at home* in extraterrestrial space, then one of the most important requirements is that they should experience the familiar attractive and constricting forces associated with the gravitational environment at sea level. It also has been the experience of early space colonists from earth, that for a space colony to succeed as a place for the continued evolution and *progress* of the human species in space, there must be at least a close approximation of the processes that occur on earth… That is what *terraforming* is about in the strictest sense.

So, for example, at a very fundamental level, these feelings of earthly *deja vu* must include the artificial sensation of gravitational forces that "feels" like those on Earth. Thus, within the low-earth orbital zone of the Solar System, where there are a great number of perturbances (because of the lingering remnants of earth's atmosphere), space engineers have found ways to recreate earth-like gravity.

One technique is to design space colonies in the form of a bicycle wheel, which rotates at a rate necessary to provide between 0.9g and 1.0g of artificial gravity on the inside of the outer ring. This produces a centrifugal force: the apparent force that is felt by an object that is moving in a curved path. The force acts outwardly, away from the rotating center of the wheel, which also is the center of gravity and attracts bodies within the wheel. Thus, the rotating wheel design of the interplanetary space colonies essentially is a way of "importing" earth's gravitational environment for these artificial habitats.

Consider that the human colonies in space have created societies that can only be described as replications of the "Garden of Eden" in the Common Memory of the species. As an illustration of this, the ultimate choice of food production systems on orbiting space colonies reflects an understanding that earthling food means much more to humans in space than simply a combination of nutrition and calories. For more than two million years, the acquisition of food – either by hunting-gathering or, through agriculture or aquaculture – has cultivated a web of human emotions that cannot be replicated by any other activity or industry.

Indeed, in the post-industrial age of the 21st century most "work" now either is done by machines and managed by computers. Consequently, modern humans are returning to the principle pastime of our ancient hunter-gatherer ancestors: the gathering and preparation of food, and the sharing of meals… as a means of acculturation. In other words, enjoying the leisure time of the ancient hunter-gathering societies… making a central pastime of the preparing and enjoying meals and conversation… bonding and enjoying the "good life" of simple societies that are blessed with all necessities of life… and some wants too: the arts, poetry, music and philosophizing… the fruits of the enhanced human mind.

The important point to recognize is that carbon-based humans continue to attach *emotional value* related to the whole range of activities surrounding the preparation of food. With the above in mind, food on space colonies is now being produced *in situ*, through hydroponics and other technologies, processed within the space colony and distributed in the markets reminiscent of the medieval societies of the Old World. Making humans in space feel at home will ultimately be accomplished by focusing the activities of these terrestrial migrants related to *food*.

Once again, like their Pleistocene and earlier Holocene ancestors, carbon-based humans in outer space devote considerable time and energy on producing food for life and food for thought and their culture, on the staff of life. Food is once again a significant focus of life events aboard the space colonies – from on-board fields to the tables. Food nurtures the organism and the soul; the arts and the artisans… Most of all, a food culture helps resolve the feelings of estrangement, homesickness and the mind-numbing boredom of existence in space.

Walter Gomez

Space Colony Food Production Systems

The plants which developed on Earth require several *inputs* to grow, regardless of where they are transplanted. The basic formula for growing plants everywhere, including in terraformed extraterrestrial places, is as follows:

$$Pf = (S + Te + L + W + CO^2 + N + Ti)$$

Where:

Pf = Plant food
S = Space
Te = Temperature
L = Light
W = Water
CO^2 = Carbon Dioxide
N = Nutrients
Ti = Time

All these inputs can be obtained – without extraordinary landing-launching events – by space colonists, by drawing from the veritable Periodic Table of Elements that exist on asteroids, comets, moons and other low-mass celestial bodies, which can be maneuvered into proximity with a space colony.

The food-production technology of choice on the space colonies is aeroponics, which is like hydroponics, except that gas, rather than water, is used as the planting medium. In aeroponics, no growing medium is used. Instead, the roots of plants are suspended or hung in a dark chamber and periodically sprayed with a nutrient-rich solution.

As the result of more than a century of scientific and engineering experiments to test the viability of growing food in space, it has been decided that the best technological paradigm is that of aeroponics. The years of experience in using these methods for producing food aboard space laboratories, also has shown that micro-gravity environments provide excellent mediums for magnesium, zinc, and iron... for any earthly plant. And indeed, through the application of precise and controlled inputs of these elements, and the use drip irrigation techniques... the aeroponic units onboard the space colonies are now producing adequate amounts and varieties of foods that are as nutritious and "tasty" as those that are remembered "at home."

However, just as important to the nutrition benefits that derive from the onboard food systems, are the beneficial aspects of photosynthesis... which are being derived more precisely than is possible on Earth, largely because of the ability to manage the rotation and orbital movement of the space colony with respect to the Sun's light. A beneficial corollary of is that oxygen is produced as a byproduct of photosynthesis, whereas the density of carbon dioxide is maintained at optimum levels by scrubbing and dispelling processes, thus maintaining a healthy ecological system.

Then there is the *creature comfort* aspect; the familiarity factor: the recognition of the psychological and social benefits that are derived from the producing "locally-grown" and "organic" earth-like foods.

And, along another dimension, the research that is being done to develop earth-like foods has engendered advances in the understanding of the cell, and has wrought technologies for engineering the cell. This greater understanding of the building blocks of carbon-based humans has also been transferred to achieving more sophistication and agility in engineering the atom and chemical elements.

Ultimately, having an onboard the space colonies has engendered an acute awareness of the ecology that exists within the space colony, and the need to maintain it at a level which creates a better "quality of life" – but also instills the knowledge, that the production of foods that are healthy, highly-nutritious and satisfying of the psychosocial needs of the space colonists, is also a highly-effective and efficient means for maintaining the physical and psychological health of the human individual and of the human population aboard space colonies. Food is truly the best medicine for humans, and for their environment.

No Human is an Island

Feeling at home, for modern Holocene humans, also means being interconnected with other humans and artificial intelligence entities. This basic need for social interaction first emerged in the Pleistocene era when families organized themselves into food acquisition systems, with communal purpose of mind. This seminal development of *group think* enabled them to devise tools and techniques, as well as communication and organizational skills to develop a successful strategy for surviving and propagating in their relevant natural environment.

During the Holocene period, when the survival strategy changed from a passive acquisition (hunting and foraging) to a proactive production (irrigated agriculture) of food, the need for more sophisticated communal systems developed. Daily interaction among individuals intensified, and so did interaction between central places… thus creating larger areas of interaction. Gradually, the social gene – which was enhanced during the period of recovery from the global reconstruction that followed the end of the last Ice Age – would be passed on to succeeding Holocene generations. The most obvious manifestation of this phenomenon would be the emergence of metropolises (central places) and their linkages.

Developing Linkages

The strategy for ensuring the survival of humanity in outer space today is somewhat like that of the Pleistocene hunting-gathering societies on Earth. In both cases, the basic unit of society consists of a small group of humans which are connected by an *esprit de* purpose. Another thing which they have in common is a penchant for opportunistic *in situ* gathering of natural resources... and, therefore, they are always traveling... and sometimes encountering other human groups, and thereby are creating linkages of various sorts.

Thus, the human colonization strategy throughout the two million years since the GHAE has involved the propagation of the species by relatively small, genetically-homogeneous, and culturally tightly-knit familial groups – even into the 21st century – to colonize the planet Earth. This paradigm continues to be relevant and efficacious today in outer space: the colonization of the extraterrestrial regions of space is still being conducted by highly-organized and tightly-knit – but much more diversified – groups of beings. These include not only CB humans, but also all the *sentient* and *intelligent* semi-organic cyborgs and non-organic SB humans.

The linkages that were manifested by foot trails and established pathways of the Pleistocenes, and the roads and highways of the Holocenes on Earth are mirrored in space by the trajectories and orbital paths of spacecraft and electromagnetic beams. It is a manifestation of the continued emphasis on developing robust linkages throughout the history of human colonization, and reflects the notion that – no matter how great the cumulative knowledge or how powerful their tools, or how elegant their skills, no node can exist alone, in splendid isolation – not if earthlings are to succeed in the overall objective of colonizing the entire Solar System.

This means that each space colony must develop and constantly cultivate effective linkages with the other colonies and with the natural nodes of the Geospace Region. To begin with, this means finding the most fuel-efficient pathways through outer space; the optimum points of insertion into the orbital "highways" and the most efficacious gravitational means of navigation for accessing the natural gravitational momentum from massive bodies.

This development of charts of the gravitational oceans is also done by silicon-based, intelligent robots which can live and work on the construction site with no concern for life support systems. Beyond the construction of the infrastructure, the SB humans also are tasked with creating the onboard the "life support" system which will enable carbon-based humans to populate the colony.

Chapter 30

Creating A Geospace Region

If one were to characterize Holocenes (*a la* Homo habilis), it might be under the rubric of *Homo architectus* (builder). This has been manifested in the construction of all manner of edifices: from pyramids to skyscrapers. The same epigenetic urge to build has been seen in the way that Holocene humans have constructed artificial spatial systems (regions) on virtually every area of the earth's surface. Now, space colonists are using the Common Experience of creating human regions on Earth to develop similar artificial spatial systems or regions in space.

Long before the first space colony was established in the 21st century CE, the Enlightened Minds of the Ionian islands had already created a mental map of the human region that would be developed in outer space, only they envisioned the space colonists to be divine beings.

A few Enlightened Minds of the time even intuited that the human region in outer space would be geocentric because the central node of this celestial human spatial system would be Earth… and all the other nodes of the system would revolve around it.

Region Building

One of the greatest gifts of knowledge from the Cosmic Mind to humans has been the knowledge of *region-building*. It is represented in mental maps of places – those that have not yet been experienced by the human brain. These highly abstract maps depict the essential nodes and linkages of a given place.

Once a virtual region has been mapped within the human mind, it sometimes is passed on to certain *mutant* human brains, where the abstraction is gradually (or suddenly) recreated as a physical model of a place. This phase of region-building can be compared to the way in which 3D printers create reality from abstract alphanumeric instructions received from a computer.

In the abstract, the Solar System is essentially a vast, but still finite *three-dimensional space*. At its most basic level, it is a field of energy and matter that behaves according to the laws of gravity. A derivative of this for human colonists in space, is that there are an almost unlimited number of potential pathways for transiting the gravitational field of the Solar System. These natural pathways appear as orbits of various angles from the elliptic and whose patterns are controlled by the nearest, most massive celestial bodies.

These potentialities are naturally exploited by artificial massive bodies (e.g., space colonies and spacecraft) in their intended travels from the Earth or from any other point of origin… to any other point in the Geospace. *Intended* is the keyword, however, because there is always a high probability that a pilgrim body will have its journey truncated by all manner of perturbations in orbit – caused by collisions with other massive bodies or degeneration from the heat produced by kinetic energy.

Energy – in the form of photon units and conveyed in waves – also travels throughout the heliosphere in multiple directions. The points of origin of energy fields are varied, some emanating from the life cycle events of the galactic stars, others from our own star, and still others as the result of atomic decay of the elements. Planets and moons of the solar system also emit energy from volcano eruptions.

[An important point with respect to these energy fluxes, plasmas and fields within the Solar System is that these can be utilized

by humans as conduits and pathways to connect the nodes of the Geospace Region. Among other things, they are pathways for conveying payloads of all kinds: everything from raw data to information; through the various phases of the electromagnetic spectrum... encompassing all the elements of the periodic table.]

Like all human regions, the *cultural* Geospace Region is constructed on the foundation of a physical field. Within this physical field, all natural phenomena are regulated by the laws of gravity, electromagnetism, and the strong and weak forces. By extension, all *cultural* phenomena are likewise subject to the mechanisms derived from these forces. Only by using human technology, can the rules of the field be managed.

Thus, the boundaries of human regions in space are broadly defined and constrained by the laws of gravity: the movement of matter and energy is regulated by the interaction of electromagnetism and gravity. And, therefore, the main challenge in building of human regions in space remains the same as on Earth: humans must devise ways to negotiate the possible – by application of technologies. In the case of the Geospace Region, every action related to its development has been basically regulated by the forces of gravity and electromagnetism and, less so, by the strong and weak forces. But when *Homo architectus* began to inhabit outer space, even the most powerful forces in the field would be harnessed, to some extent, by human technologies.

[Whether it involves development of propulsion systems; the creation of artificial materials; the development of artificial life support systems; or the movement of all manner of tangible and ethereal phenomena throughout the region... it is now obvious that Earth has been an incubator, which has prepared humans for carrying out the mission of colonizing the Solar System.]

By the year 2031 CE, humans had established the prototype of the Geospace Region on Earth; and in creating a new human region in space, the space colonists were drawing from the Common Experience of both the Pleistocene and Holocene cultures. As a result, there are many similarities between the Pleistocene self-contained colonies and the self-contained space colonies. Thus, for instance, humans are once again colonizing the rest of the Solar System – one colony at a time – to form one constellation after another. This is reflective of the success of the basic strategy of colonization that has enabled earthlings to survive and thrive in unknown natural environments. So, once again, there is the familiar pattern of developing and propagating colonial nodes, and the growing of the linkages that bind them into a functioning human spatial system in outer space.

Earth, Mars and Jupiter will be the primary nodes in this emerging human region of the Solar System. The relative importance of these nodes in the human region will not be determined by the relative magnitude of mass and gravitational friction, but other "human" factors, such as comparative advantage in terms of accessibility to natural resources, relative distance and location with respect to other bodies, both natural and artificial, or the number of *visits* that are made to them by earthlings or residents of other colonies; that is, magnitude of human traffic between nodes.

Like any human region, the Geospace Region requires energy to maintain its *viability and efficacy*. Overall, the Sun is the ultimate source of energy for the orbiting space colonies. In the case of the planetary nodes, however, the effective amount of solar radiation is often diluted by magnetic fields or atmospheres.

The Earth is the only example of this, for the present. It has several layers of filtration: the magnetosphere and the various fields of charged particles within it, as well as the various stratified layers of atmospheric clouds and dust to shield the surface. Mars, on the other hand, is unable to maintain an effective atmosphere, but it does generate significant mitigating fields because of its global dust storms. Jupiter is protected by the layers of gases that cover its surface.

Chapter 31

Governing the Geospace Region

Drawing from more than 4,000 years of Common Experience on earth, planners decided to follow the model of the city-state in in developing the space colonies. There are several salient parallels between the situational relationship between Mediterranean Sea and the Hellenic city-states, and the interplanetary medium and the orbiting space colonies. Like the ancient Greek city-states, each space colony is organized as a self-governing political unit, within the framework of geo-international law.

Recognizing this commonality, the nations of Earth have agreed that the risk of the survival of humanity on the home planet had reached a stage where it was necessary to finally fully implement the United Nations as the *de facto*, as well as *de juris* central governing body on Earth.

As such, the UN would be the controlling political agency for planning and organizing the human colonization of the rest of the Solar System. There would be a separate Director of Space Colonization, an executive function that would answer to a legislative body which would be comprised of representatives from each of the orbiting space and planetary space constellations. A judiciary branch would dispense justice and assure fidelity to the founding documents and derivative laws and regulations.

Founding Documents

The colonization of the extraterrestrial Solar System officially began on December 31, 2020 with the approval of the United Nations of the Law of the Interplanetary Medium by all the member nations of the United Nations. Otherwise known as the Convention on Space Colonization, it would be the document that would guide space colonization by the earthlings.

Following the paradigm of the ancient Greek colonies and later, the English Colonies in America, each space colony is granted a charter from the United Nations, which lays out the rights and obligations of the colonies with respect to each other, and with the United Nations.

A charter establishes strategic objectives for every phase of the mission to establish human space colonies throughout the Solar System. Underlying all these principles is the overriding imperative of human survival and propagation everywhere. Inherent in this is the understanding that the tactics of *dispersal* and *diversity* are essential for ensuring the survival of humans in every variation of physical environment – thus increasing the probabilities of success in surviving and thriving throughout interplanetary space and on the other planets, moons, and asteroids of the Solar System.

In practice, this has meant that each colony must be self-sufficient, with the capability for autonomous operation, but also must be equipped to function as an integral node of a regional *constellation system* of space colonies.

To this end:

Flexibility of scale is a basic element of the human space colonization effort. It derives from the lessons contained in the Common Memory of the existential events that threatened to extinguish the human species on Earth during the closing stages of the Pleistocene period.

Alternating dispersal and re-integration is another basic element in the survival for the human species in outer space. The implementation of this seen in the diversity of demographic and genetic composition of the human populations within each colony. Clearly there is a purposeful attempt to create highly-diverse *genetic colonies*, in which the DNA mix is designed to achieve desired adaptations for *permanent* human residence in space. The underlying hypothesis is that a greater number of potential combinations of genes provides each colony and constellation with a greater capability for developing the optimum human entity for colonizing every environmental niche in the Solar System.

Geospace Political Systems

> *"They believe that all of the nations of the world, for realistic as well as spiritual reasons, must come to the abandonment of the use of force. Since no future peace can be maintained if land, sea or air armaments continue to be employed by nations which threaten, or may threaten, aggression outside of their frontiers, they believe, pending the establishment of a wider and permanent system of general security, that the disarmament of such nations is essential. They will likewise aid and encourage all other practicable measures which will lighten for peace-loving peoples the crushing burden of armaments."*
>
> (UN.ORG)

Peaceful Consensus and Cooperation is the underlying ethos of this first extraterrestrial human society. For the first time in human history, a society has internalized the imperative to ensure the survival of the human species through cooperation and peaceful resolution of competing interests, instead of through conflict. The ethos of the jungle and the savanna, the clan and the nation, which encourages a competitive, a zero-sum contest for tribal survival has been replaced by one of cooperation and consensus… whose aim would be the survival of the species.

The cooperation paradigm has become the underlying sense in the preparation of humans that are sent out to colonize outer space. So, one of the most important aspects of the space colonization program is the genetic and psycho-social engineering of presumptive colonists, prior to sending them to live and work in space… whether in orbit or on the surface of a celestial body.

It is a purposeful attempt to develop in humans a certain *mindset* – one which naturally permeates the space society with the values of social cooperation, rather than competition between cliques within colonies, and between colonies. The truth of the matter is that a spirit of coordination, which promotes the objectives of colonization – as opposed to debilitating competition – has proven to be the most efficacious motivator for achieving the ultimate objective of colonizing the Solar System.

One of the most important reasons why this is now possible, is that there is no need for aggressive competition in extraterrestrial space, where all the natural resources that are needed to construct and maintain a human colony are both ubiquitous and accessible. They are virtually *free* in economic terms and, therefore, not a source of inter-colonial conflict. H2O, for instance, is everywhere available – given the present level of technology – as a source of both water and oxygen to sustain all manner of *earth life-forms* in the colonies; and it is readily convertible as a fuel source as well. And, having gained *familiarity* with the gene and the cell, as well as the atom… human space colonists no longer face famine or disease; there are no natural or cultural destabilizing forces to create wars of desperation.

There is no religion or mythology… no tribalism or nationalism… no ideology or philosophy to engender destabilizing or destructive competition within the societies of the colonies and between the colonies. The space societies are based on the internalization of the imperative for the survival of the individual and the species. And, all *individuals* in space societies, whether carbon-based or silicon-based, are equally valued.

Thus, the humans who are colonizing the rest of the Solar System have been engineered to seek peaceful resolutions of conflicts. The carbon-based humans have been genetically-engineered to be less aggressive by removing the vestiges of the primate "fight or flight" tendencies, and the silicon-based humans are similarly programmed. The upshot is that social conflict is now almost always resolved by the powerful instinct of empathy and the resultant effectiveness of communication. But this is not to say that there are no conflicting issues in space societies – only that effective mechanisms for debate and consensus have been put into place.

Founding Princples of Geospace Colonies

1. The Solar System is to be used for peaceful purposes only.
2. Humans are only the stewards of the whole Solar System; no territorial sovereignty is ascribed to any human subgroup.
3. The United Nations is the executive entity for administrating policy in space.
4. The United Nations Peacekeeping Force maintains security in space.
5. The General Assembly of Colonies engages in continuing consultations to make policy decisions.
6. The Geospace Court of Justice provides a forum for peaceful settlement of disputes in Geospace.

Systems of Government

The system of governance that has been adopted for the Geospace Federation is the product of the Great Human Awakening Event and of the ongoing body of programmed instructions from the Cosmic Mind that have guided the entire genetic and epigenetic development of the human species. It is still influenced by the experiences of the hunting-gathering societies that formed soon after the GHAE and those of the ensuing Holocene agrarian, industrial and post-industrial societies.

Another contribution to the system of governance of Geospace Federation is the experience with various forms of government that developed in the city-states of the Eastern Mediterranean during the early Holocene Period. Like the space colonies, the city-states of the Greek and Roman Empires were essentially politico-social incubators in which Holocene humans developed various forms of governance… each of which was essentially a manifestation of various views of nature and the special and general responses of societies to the challenges of not only surviving, but also thriving as a species.

A basic dynamic that is occurring on the space colonies is the continuing adaptation and evolution by the colonial societies in response to the serial challenges and opportunities that arise. Carbon-based humans are continuously evolving; and the cyborgs and silicon-based entities are also being continuously modified and customized for life in space. In all these cases, individuals are aspiring to attain what it is to be *human*. Indeed, the diverse society that is developing on the space colonies mirrors, in many ways, the diverse societies of the ancient Hellenic colonies, which emerged all along the littoral of the Mediterranean Sea about 3,000 to 2,000 years ago. In the latter case, the diversity of the population was manifested by elements of ethnicity and station.

The reconfiguration of space society and the remixing of the human genome also are reminiscent of the resettlement of the survivors of the diaspora and recollection of human genes that followed the cataclysmic reshaping of the earth's surface, because of global warming and consequent global flooding. It is evident that space colonists have drawn from the Holocene planetary recolonization experience.

Consider that the fundamental and universal objective of human societies is to provide for individual sustenance and group viability. In this context, all human central places are, at their core, economic systems. But human centers, regardless of their stated purpose, also are places where groups of people also are continuously interacting on many other social dimensions – each of which requires a set of rules of the *field*; to ensure that the central place is always operating optimally. Obviously, then, these rules govern all forms of interactions between individuals, and between formal and informal groups; and all must adhere to the principles of *equality and justice* system, to ensure that the colony is moving towards its ultimate mission and its derived objectives.

Rules of the Field

Rules of the field have been implemented in all the space colonies; they are recognized in the statement of purpose in governance, which guides the day-to-day functioning of the colony. But it is the concept of *situational intervention* that governs the exceptional situation, which requires a "managerial intervention" by a controlling entity. Both the routine and the exceptional is taken into consideration by the rules of the field, as expressed by Space Law.

The model of governance in space more generally is derived from the human experience on Earth and during the period of space exploration. Much of it draws from the history of exploration and science work in Antarctica, which was carried out within a political framework by which the normal competitive stance of nation-states was essentially abandoned; so that unnecessary rivalry and conflict could be avoided on that continent. One of the relevant concepts that emerged from the experience of Antarctica was the notion that the natural resources of that region belong to the human species, in common, and that no nation or non-governmental group can declare sovereignty over any territory.

Utilization of natural resources is a part of space law which has been developing since the advent of space colonization. Fortunately, scarcity or accessibility are not significant issues; there are vast quantities of all needed resources; and they are readily accessible to all the space colonies in the form of asteroids, comets, and meteors… all of which are rich in water and minerals. The consequence of all this has been the generally-accepted principle in dealing with natural environment during the process of extraction of the resources; *conservation* of the natural environment has emerged as an ideal in space. And, consensus enlightened by the human experience on Earth has created the imperative for leaving the slightest possible "footprint" at the extraction site.

All this can be characterized as humans in space reacting to the hard lessons they learned during the Industrial Age which began in the 19th century CE. It involved a greater use of fossil fuels for energy, which then increased the carbon-dioxide load in the atmosphere and in the oceans. More generally, the humans in space resolved to treat the natural environment with greater care; viewing it as another chance to reconsider the notion of "separate and superior."

Chapter 32

The Making of Space Societies

Space colonies can remain completely self-sustaining as closed systems for one or two generations, but for long-term sustainability, there must be on-going interaction among the various nodes of the Geospace Region. One reason for this is that relatively isolated groups of any species (including humans) tend to drift towards extirpation and even extinction, if there is not some mechanism for promoting the continuous intermingling of external genetic material, to maintain the level of diversity that ensures the long-term viability of the human species.

The model for developing small, self-sustaining planetary colonies actually developed during the transition from the Pleistocene to the Holocene Period, when a new variety of humans, Homo sapiens, finally took the steps to convert from a global society of nomadic wanderers that lived off the land, as it was presented, to proactive "developers" of central places.

Pleistocene groups were continuously-moving and interacting with other population nodes during most of their 2 million-year tenure on Earth. Even though each colony usually numbered only about 10 or 12 persons, their nomadic hunting and gathering lifestyle resulted in significant interaction with the other nearest group of humans. Also, the boundaries of their resource regions often overlapped and caused constant encounters between groups of humans, which were occasions for trading of tools, artifacts, ideas, "mirror" learning, and no doubt, the exchange of genetic material through sexual encounters. Thus, Pleistocene humans and Archaic Homo sapiens solved the problem of DNA exchange by constantly walking from one place to another and occasionally bumping into other clans to *deliver* and *exchange DNA samples*.

Holocene humans took a different approach to the challenge; they stayed "in place," rather than moving from one food resource site to another. Under the new "settlement" paradigm, they would find a place which contained more resources: more water and fertile soils. It occurred to them that the best evidence of the comparative advantage of a place is the fact that it already has a more active natural ecology. So, they proceeded to created places with comparative advantages that were nurtured by human technology, and they also cultivated economic linkages among the other nodes of their spatial systems.

The Carbon-Based Holocenes

The Holocene humans revolutionized the way humans moved about the surface of the Earth and they also increased the intensity of social exchange in the process. Instead of following the erstwhile nomadic hunting-gatherer and coastal fishing-gathering lifestyle, a significant proportion of the human population switched to a settled, intensive agricultural and animal domestication way of life. On the other hand, there continued to be large populations of nomadic humans who traveled from one settlement to another, where they exchanged furs and stones for manufactured tools and artifacts, and genetic material. This blending of nomadic and settled ways of life proved to be beneficial to the advancement of the Holocene societies.

Humans have always been driven by a *common subconscious urge* to engage in social interaction through a variety of mechanisms. One of the most powerful of these is the apparent instinctive desire to trade goods and services – first through bartering and later through the exchange of analog items of value. Once the practice of economic exchange took hold, all that was needed to maintain such systems of economic interaction was the common application of *value* to whatever was being exchanged – goods and information for example.

Humans have continued to pursue a stay-in-place strategy in utilizing the resources during the thousands of millennia since the GHAE. As they have developed larger concentrations of populations in fewer central places, they also have continued to create new technologies of transportation and communication to encourage personal interaction; the most important of which has been the exchange of DNA to maintain the vitality of the society; and to avoid its extinction.

So now, in the age of space colonization, there is once again a demographic situation in which relatively small groups of humans are living at great distances from each other. Unlike on Earth, physical intercourse as a method for exchanging genetic materials is often not a practical option in outer space. However, the science and technologies related to human reproduction – such as artificial insemination and incubation, and artificial recombination of DNA in a laboratory – to deal with the problem of maintaining a vigorous, diverse human genome in the Geospace Region.

Consider that these technologies now make it possible to achieve two major objectives in the quest to develop populations of *Homo spaciens*; which are tailored to the various natural environments of the interplanetary medium and those of the various planetary domains. Thus, humans can now be genetically tailored to deal with the challenges presented by variations in gravitational and oxygen environments, as well as temperature and ambient pressure environments. Furthermore, some of these tailored adaptations can be made in practical *real-time,* as opposed to the thousands of generations it took for these to occur on Earth during the Pleistocene and Holocene periods of previous human history.

Challenges to survival of the human species also has been met by micro-engineering, to minimize the size and weight (mass) of "genetic payload," so that it can be efficaciously transported to the remotest human space colony. As it happens, it no longer is it necessary to transport an entire human body as a vessel for the DNA-RNA payload. This represents an existential leap forward in the way the human species propagates; it equals in significance the definitive break between humans and nature, which occurred more than a million years earlier, when Pleistocenes began to see themselves as separate and superior to the natural world and all it contained.

Now, earthlings in space can replicate themselves by artificial means, whether they are carbon-based humans who engineer the DNA-RNA materials to create a new being that can withstand the radiation and other extreme conditions in the interplanetary medium, or the crushing atmospheric pressure of Venus and the Gas Giants… or to be able to operate in the open for longer periods of time on the polar caps and deserts of Mars, with less application of artificial technologies. Robots also now are imbued with the intelligence to replicate parts or all their systems… the same is true of cyborgs.

The Cyborgs

The term, "cyborg" originated in the realm of science fiction, but as so often has happened in the age of space exploration and colonization: scientific reality has caught up with intuitive conceptualization of the mind. So now, the cyborg has become an integral part of the human resources of space colonies. The *bionic human*, which refers to a normal human organism, whose innate abilities are enhanced by transplantation of enhanced organic and non-organic components, as well as genetic engineering.

The creation of a *human being* that is not all flesh and blood has been relatively easy to create, given the rapid advances of the past half-century in the technologies of genetic engineering, organ transplantation and in the integration of artificial prosthetics to the basic human organism. Harder has been the effort to gain acceptance from the carbon-based and silicon-based humans on the space colonies. Part of the reason for this is the fact that cyborgs do not contribute obviously to the mix of current human resources on space colonies. Rather, cyborgs are manifestations of an ongoing program to develop a human entity that is more like a carbon-based human "flesh and blood", but also possesses the innate capabilities of the silicon-based human survivability in the extreme environments of the Solar System.

But, though the integration of the cyborg into the overall human populations was a difficult to accept by earthly humans, when it was first introduced in the latter half of the 20th century... ultimately, as the benefits of *smart* prosthetics came to be appreciated and commonly accepted as simply another expression of epigenetic human evolution, the cyborg became just another variant of the human species.

The iconic example of the making of a cyborg human in the 21st century has been the gradual metamorphosis of the scientists and futurist, Stephen Hawking – from pure flesh and blood – to a progressive amalgam of organic and non-organic components. So, as the progressive ALS neurological disease continues to attack his muscles and the connections between them and his brain – Hawking has been fitted with one prosthetic device or electromechanical system after another, to enable him to maintain a somewhat erect posture in his *smart* wheelchair, and to continue his work as a scientist, author, and a translator of the universe to other humans.

Electronics and sophisticated computer software also have enabled Professor Hawking to continue oral communication by emitting computer-generated words. And, when he wants to interact with his computer, he simply emits electrochemical signals from his brain, which interfaces directly with an advanced computer system. He has become, in short, a cyborg entity whose carbon-based brain retains its considerable powers of logic and reasoning, and interacts electro-mechanically with the artificial silicon-based parts of his body.

So, cyborgs are now an integral member of the human resources on a space colony. They often are the executive entity of the colony, because they have the mind of the carbon-based human and the immunity to hazardous conditions, as well as the powers of the silicon-based robot. In the future, the cyborg will become an important member of the "human" resources of space colonies.

The Controlling Mind

The Human Mind has become the controlling agency in the colonization of the Solar System by earthlings…

Humans first received the gift of sentience during their incubation period in Africa, as part of the GHAE. Since then, the human sense of internal and external awareness has continued to grow, as the senses of the brain and of the mind have continuously brought light to the darkness of incognito of the universe. With these powers, the human mind has emerged as the agency of the Cosmic Mind, and continues to be the prime mover in the colonization of the Solar System. It receives instructions from the Cosmic Mind and implements them as a kind of blueprint.

By the time the first planetary migrations of colonization began, approximately 1.8 million years ago, humans were already using the higher consciousness of the human mind to rationalize their dominion over the other animals and the plants on earth. This was the first step in defining the essential separation between themselves and every other aspect of the physical world; which they did by constructing a virtual construct of reality consisting of three basic entities: the gods, their human derivatives, and the physical world around them. From this syllogism, they concluded that they were granted dominion over the natural world by the gods.

Another re-examination of the relationship between humans and the gods; and between humans and the physical world… occurred again in the 21st century, as modern Holocenes developed the power to replicate the human organism… its brain, organs and exoskeleton… with the essential building blocks of the atom and the cell. Once again, a grand rationalization of the human mind provided a logical construct which empowered humans to proceed with the engineering or manufacturing cells and neurons, and the other aspects of the natural human anatomy and physiology.

A third revaluation of the relative place of carbon-based humans in the cosmos occurred as SB humans were on the verge of developing a virtual human brain… sparking a debate about how much of the essential elements of

CB human intelligence should be granted to "robots", especially the humanoid robots. This debate now is being framed in terms of *Artificial Intelligence (AI)* and it has occurred in several arenas. Two of the most important of these are cast in the logical syntax of religion/philosophy, or that of reason/pragmatism. Today, the latter predominates in the space colonies.

Human have now accepted the notion of *silicon-based* (SB) humans. And, in the year 2020 CE, the first constellation of orbiting space colonies came on-line, as the first complements of 10,000 CB colonists arrived at their assigned space colonies... and were received on the colonies by teams of SB humans who had been working to construct and to bring each colony into operation. Even after the arrival of the CB humans, the SBs continue to perform the bulk of operations and maintenance activities on the space colony. Any activity that is too dangerous, psychologically numbing, or troubling for CB humans is assigned to the SBs.

And, as it has turned out in practice, the CB humans have had relatively few problems in accepting the notion of co-equal non-carbon humans as part of the societies in space. Thus, many space colony societies include both CB and SB scientists and engineers; both continue to do the work of space exploration throughout the Solar System. And, space probes, planetary rovers, and onboard computers have become familiar actors in space too.

Robonaut and "HAL" are iconic examples of the changing role of silicon-based humans to the overall mission of colonization of space and the bodies that inhabit it. They are silicon-based, artificially-intelligent beings, but they take a variety of forms, depending on the specific function they fulfill on a space colony. Or, they can take on a thoroughly "human" form and present an affect that is comforting to carbon-based humans; in the manner of the Chewbacca character of the Star Wars franchise, the comfortable crewmember of the space colony.

Chapter 33

Human Resources on Space Colonies

The *human* resource systems within each space colony must be *diverse* in many ways, because they must operate in various environments, and because they are assigned such a wide variety of tasks. It is also the case that carbon-based humans are unable to operate in many problematic space environments, even within the protective cocoon of modern spacecraft. Therefore, silicon-based humans and cyborgs are an increasingly important component within the human resource component of space colonies, in orbit or on the surface of celestial bodies; and the proportion of silicon-based humans aboard space colonies that are being deployed in the outer reaches of the Solar System – where the logistical lines a more tenuous than ever – is constantly modulated to respond to changing missions and varying circumstances. And, so, carbon-based humans successfully live, work and recreate alongside SB humans and cyborgs.

From the very beginning of space colonization, robotic workers have been involved in the preliminary construction and the continuing operations of the extraterrestrial colonies, even after the arrival of the carbon-based colonists. Thus, the new carbon-based colonists integrate into existing silicon-based societies, instead of replacing them. And, even as the new immigrants are being integrated into the existing silicon-based colonies, the overall society aboard the space colony continues to undergo many changes in the makeup of its human resources.

The Worker Bots

Like the Holocenes have done on Earth with their work animals and machines, SBs deploy artificial entities to do work and perform specialized tasks. Often, they are used in situations where the peculiar *architecture* of the so-called "bots" is best suited for a specific task. Bots come in many sizes and forms. There are those that look and move like arachnid or insect form of life on Earth. These silicon-based entities are designed to use *swarm* techniques to execute their assigned tasks and, therefore are programmed with the specific cognitive and reasoning capabilities that are necessary to detect and solve unforeseen problems… by sensing, evaluating and correcting, all in real-time.

So, their main function is to maintain the outer "epidermis" of the space colony; which itself contains embedded sensors to constantly detect, analyze and auto-correct for changing external environmental issues. At the same time, they absorb data from the interior environment of the colony; evaluate the situation and, if necessary, actuate changes in the material composition or neural activity of the *skin* of the colony. These modifications enable the colony to respond to variations in the *threat environment* or the *opportunity environment*, to ensure *ops normal* conditions. Meanwhile, they continue to carry out the routine tasks of ongoing maintenance, and the occasional updates and additions to the superstructure and the internal systems.

Thus, the normal division of labor assigns virtually all the "white collar," repetitive "left-brain" work, which is traditionally associated with the ancient scribes and the modern "bean-counters" and computer gurus. But the division of labor between *smart* and *dumb* robots is a shifting boundary. *Smart* robots normally operate autonomously, as per prewritten software coded instructions that include "Boolean" decision-making algorithms… without any real-time control by other entities. Some of them are even capable of "updating" their own programming instructions in response to anomalies and unforeseen circumstances.

These often are supervisory workers who control the actions of the so-called *dumb bots,* or interact directly with the onboard computers. They, for instance, control extra-vehicular and infra-vehicular which require real-time and instantaneous sensory-logic-response operations in the face of the extreme life-threatening occurrences. But it is the SB humans that do all *extravehicular* space work; mainly because they can reside and work in alien environments *au natural*... and as easily and effectively as the early Pleistocenes, who were virtually naked and equipped with only a stick or a rock.

And even today, the SB *humans* continue to the task-level work on the construction sites of the new space colonies; those within the Cislunar Subregion, as well as the Martian, Jovian, and the Plutonian celestial bodies. These construction workers create the basic infrastructure, including: the *cosmodromes* and the refueling stations, the repowering and refitting bases; the communication relay points; and all the *caches* and depots that are needed to maintain these linkages for the constellations of colony *systems*...

Certain CB-SB human hybrids – the so-called *cyborgs* (cybernetic organisms) – have found a special niche within the worker population on space colonies. These are carbon-based entities that have undergone genetic engineering and/or artificial enhancement of the basic organism... to enable them to operate in environments which are too dangerous to the unenhanced human body or mind; or even just too boring for unenhanced CB entities. And yet, they continue to *feel* the emotions of carbon-based humans.

For these reasons and others, the cyborgs have become a prime platform for conducting research and development in creating an enhanced hybrid carbon/silicon-based humans, with the best features of each, to work along-side the silicon-based humans in preparing planets, such as Mars and the moons of the Jovian planets for eventual permanent colonization by carbon-based humans.

In the years prior to the launch of the first constellation of orbiting space colonies, the division of labor on Earth had already presaged that aspect of societies in space. The first generation of SB humans (already known as robots or bots) had assumed virtually all the tasks of the muscle and most of those of the left-hemisphere of the brain. Some artificial intelligencias were even beginning to dabble in the creative arena which is the domain of the right side of the brain. And, smart machines and integrated computer systems increasingly took over functions of supervision and of middle management.

But, ultimately, it has been the carbon-based, Enlightened Minds that have assumed the role of "philosopher-kings" of a sort in the development of the space societies. They also are fulfilling the role of a special conduit between the Cosmic Mind and the minds of the humans in space. Looking forward, these will be the human entities that will foresee and guide the trajectories of the continuing space colonization by earthlings.

Geospace 2060

Human Resources On Space Colonies

CARBON-BASED HUMAN

SILICON-BASED HUMAN

CYBORG

BOT

SO NOW – AS IN THE ANCIENT SOCIETIES OF EGYPT AND THE HELLENIC WORLD – THE CB HUMANS OF THE SPACE COLONIES HAVE BECOME THE PHILOSOPHERS AND ORACLES; "IDLE HANDS" AS IT WERE, WHO INTUIT THE "UNSEEN" PHYSICAL WORLD AND CREATE MODELS OF HUMAN INSTITUTIONS AND SOCIETIES THAT ARE OPTIMALLY ATTUNED TO DEAL WITH THE A PRIORI FUTURE...

Consider that the realm of the human mind is now that of the *dreamer* and the *intuitive learner*. In erstwhile Holocene societies, these individuals have been referred to as Enlightened Minds who have reached a level of consciousness – either by genetic mutation or through epigenetic awakening – such that they can communicate directly with the Cosmic Mind and the Common Human Mind. In space, all the CB humans are endowed with this level of enlightment.

Throughout the Holocene Period, there have been Enlightened Minds that have engaged in *contemplation* as their main activity, rather than doing the mundane tasks related to the physical survival and propagation of the human species. These have been known variously as philosophers and priests; shamans and oracles; prophets and futurists; and savants and geniuses.

The earliest Enlightened Minds left only a scant record of their work – through symbols and images inscribed on the walls of caves, or imprinted on stone memoranda, or some other non-decaying media. Later generations of EMs have left memorialized their works on mud tablets, papyrus, and other portable media...

Sometimes their thoughts have been memorialized in the songs and stories of their descendants. These are the memories of the right-brain that have been embedded in the genetic material of humans. Indeed, most of what is known of these earlier Holocene Enlightened Minds today has been archived in the genome of the thousands of generations of humans on planet Earth.

Contemplation, Beauty and Empathy also have emerged as the exclusive pursuits of the carbon-based humans on the near-Earth space colonies; however, in the newer, more distant space colonies, the preponderant silicon-based humans are also engaging in the "arts" and the other creative outputs of the right-brain.

Thus, in the tradition of the schools of philosophy and lyceums of the ancient world of the Holocene period, and the later universities, salons, coffee houses and think tanks of the modern era... both the CB and a few SB humans of the space colonies now devote their time and energies to the work of the mind, and have left the traditional work of the brain to *intelligent*, left-brain SB humans, and have assigned the "labor of the hand" to robots and bots that operate under the command of computer programmed instructions.

And so, it has come to be that the sentient and autonomously intelligent CB and SB humans represent a new variety of human – *Homo spaciens.* This appellation reflects the technological reality that in the 21st century CE, carbon-based cells and the atom have become essentially "fungible," in the sense that genetic engineers and chemical engineers have reached the point where they can utilize elements, atoms and the components of the biological cell to construct sentient and intelligent beings to carry out the work of cosmic colonization, nearly without regards to extremes of temperature, pressure or radiation.

These transcendent earthlings will also be the leaders in the continuing colonization of the Solar System, as they develop the virtual pre-reality, imagery, models, and algorithms for colonizing the Milky Way galaxy and the rest of the Universe...

Challenges and Opportunities

As to the variability of natural-cultural environments along the space-time continuum, just consider the many varieties of *micro-environments* that occur within interplanetary space and on the various celestial bodies that dwell within it. Then factor in the occasional tsunami-like disruption in the gravitational field caused by collisions between celestial bodies and acute disruptions on the galactic stars and the Sun… routine minor perturbations and turbulences that derive from the routine movements of bodies, and the currents and whirlpools within the heliographic gravitational field…

So, many transient disturbances can occur during the voyages of the celestial bodies and spacecraft through the heliosphere; they emanate from cyclic and variable fluxes of light and energy from the Sun; the latent bursts of energy from galactic stars… and from the restlessness of immature bodies… and from the progeny of fragmented bodies… or from the decay of elements that were created in the aftermath of the Big Bang event. And then, adding to the challenges of navigation in space, there are the Trojans and other "orphans" who wander about the heliosphere, unattached to any of the more substantial members of the heliosphere. They all contribute to the long-count and immediate variations in the space through which spacecraft travel and orbit, and they contribute to the uncertainties, amongst the myriad of variables, which defy precise description *in toto*, and which can only be managed within a given micro-environment, as it occurs.

THIS DIVERSITY IN THE NATURAL SPACESCAPE PRO-
DUCES AN EQUAL DIVERSITY IN THE CULTURAL SPAC-
ESCAPE OF THE SOLAR SYSTEM...

Because the human space colonies face a great diversity of natural environments, they are bound to develop an equally diverse set of cultural responses to the vagaries of the heliosphere. One of the most important of these has been to develop a variety of artificial genetic and technological responses which differ according to the specific environment in which a space colony is being deployed. That is why the colonies of the Cislunar Subregion are different than those of the Martian Subregion, of the Jovian Subregion, and those of the Main Asteroid Belt and the Kuiper Belt – why each of these is different from the others. The reality is that the variable distributions of matter and energy produce different natural environments throughout the time-space field of the heliosphere.

Resource Accessibility

The space colonies generally have effective access to asteroids and comets, which are essentially orbiting "tables of elements" and natural caches of raw materials. Many of these chunks of planetary matter are also *economically accessible* sources of water, volatiles, essential minerals, as well as sources of rock and soil materials for use as radiation shielding.

As indicated earlier, a guiding imperative that drives the colonization of the Solar System, is that each colony must be self-sufficient. Among other things, this means that wherever possible, a colony should develop the capabilities for *living off the land*, by utilizing the natural resources that are available and accessible within the *local place*; that is, from nearby asteroids and comets.

> *[Consider that the Pleistocenes also had access to water and food wherever they traveled throughout the Tropic of Cancer Region… and they did not have to worry about generating oxygen, nor did they have to create an artificial gravitational milieu or an artificial magnetosphere. By contrast space colonists must create all these elements of human life wherever they go in the Solar System.]*

Some planets and moons offer environmental niches, which are relatively manageable as carbon-based habitats; but only with significant inputs of technology. On the other hand, the quasi-void that lies between the massive bodies throughout the Sun's gravitational field is always antithetical to carbon-based humans, especially when they venture outside artificially terraformed capsules.

Only SB humans and their bots can exist *au natural* everywhere in space and on all celestial bodies, even on the crushing and toxic atmospheres of the Gas Giant planets, but even they will deteriorate over the long-count of residence in those environments. It follows, then, that earthling space colonies ultimately must always be enclosed *terra environments* – at least wherever carbon-based humans happen to reside, or in sometime in the distant future, when genetically-engineered cyborg humans can survive in any environment of the Solar System – regardless of how *alien* it may seem today.

This will become even more significant when earthlings begin the long-term project of colonizing the planet Mars and its moons, in tandem with the ongoing operations of terraforming. Plans are being developed to begin this effort by the year 2070 CE.

Chapter 34

Geospace Economics

The macro-economic system that has been adopted in Geospace follows the paradigm of *mercantilism*… a spatial and functional system previously used by European nation-states during their conquest of the Americas, beginning in the 16th century and continuing well into the 20th century CE. The metropolitan nodes of that system included the dominant nation-states of Europe and their subservient colonies in the New World.

In Geospace, a similar economic spatial system has been developing again; only this time, the metropolitan node is Earth and the subservient colonies are the space colonies. Once again, a major function the interplanetary economic system is to finance the costs of deploying and maintenance of the space colonies by the underwriters – at least until the colonies can become self-sustaining. Once again, the value derived from the raw materials in the *new territories* is being used to add to reimburse costs and to add to the wealth of the colonizing nations on Earth.

So, in the case of the Geospace colonies, the beginning relationship between the metropolitan node (Earth) and the colonies has been essentially that of a startup venture, in which the cost of construction, operation and maintenance of the space colonies and the underlying infrastructure is initially being borne by the nation-states and non-governmental organizations on Earth. But, it is expected that by the year 2070 CE, when the terraforming and colonization of Mars begins, the "domestic" economy of the orbiting space colonies will have matured, and will therefore, be able to bear the cost of the colonization of the Red Planet.

Infrastructure

The creation of a comprehensive Geospace transportation and communications infrastructure to connect the nodes another important objective in maximizing the effectiveness of the colonization effort. To this end, the national space agencies had already done much of the preliminary work of space colonization, as they carried out space exploration. So, by the time the first constellation of space colonies was being deployed, there was already a functioning network of established orbital paths, along which people, goods, services were conveyed – to provide an increasing viability to a developing human region in space.

The infrastructure for transportation and communication has been constructed in tandem. That is, the orbiting relay stations for radio and laser communications are often co-located with transportation depots which provide transiting spacecraft with both operational and "drydock" support as they ply the space lanes to other colonies or to the more distant planets of the Solar System.

Equally important, the *soft* infrastructure that was being developed during the period of space exploration to move information throughout the constellation of colonies and auxiliary bases in outer space, and which integrated them with bases on Earth. Just consider the millions of lines of programming code that have gone into constructing the various algorithms for entering desired primary and secondary orbits… as well as the millions of lines of programming of on-board computers to respond to planned and ad hoc navigational changes in the enroute environments of a spacecraft during a mission.

Then there are the new data and algorithms that are being added to the library, as the space missions of exploration and colonization continue. Among such memoranda, there are the compendia of algorithms dealing with maneuvers and other specialized actions for carrying out all sorts of space missions.

An example is the so-called *slingshot maneuver*, which has become a routine technique for utilizing the gravity of celestial bodies to provide boost in propulsion force to a spacecraft in its journey through the Solar System. Many other elaborate maneuvers continue to be developed, to enhance the effectiveness and efficiency of space navigation.

Self-Sufficiency

Under the rubric of *earning one's keep*, it has been determined that the space colonies can best contribute to the efficacy of the overall macro-economic system by producing raw materials – as input to processing and manufacturing by the internal colonial manufacturing sector; and concurrently, as input to the metropolitan economy on Earth.

The rationale for establishing this dual-track primary sector is to enable the metropolitan node (Earth) to recoup the costs of constructing the space colony system, and to add value to the existing economy on the home planet: to become an effective node within the Geospace Economy, a space colony first must become self-sufficient; and to produce the food, materials and artifacts of a modern society on the space colonies.

Fortunately, hydrogen and oxygen is plentiful throughout the Solar System. Asteroids and comets contain molecules of water (H_2O), as do some planets and moons, and from this water, oxygen and hydrogen can be extracted; the latter for use as a fuel for propulsion and other purposes. Metals and minerals are also available for manufacturing everything that is needed to maintain the existing space colony, while constructing new space colonies.

Thus, through the process of replication and propagation... humans are colonizing the entire Solar System... in units of from 10,000 to 30,000 colonists. These populations contain the optimum number of earthlings to operate and maintain a space colony, but aside from pure numbers, the demographics of the population ensures that the optimum size and composition of genetic material is available to maintain the viability of the species throughout the Solar System. Genetic diversity is also important to ensure each colony can adapt to the local micro-environment in which it operates.

There is diversity in the workforce within the colony too. Thus, most of the manual and routine left-brain work – such as monitoring operational systems and documenting routine maintenance to all internal and external system – is done by SB humans and their bots. This is task-oriented work that is normally associated with repair and maintenance and of routine affairs of housekeeping of a space colony and its equipment. While some of the right-brain creative work also can be done by artificially-intelligent entities, most of the creative planning and designing activities are done by CB humans. So, the work that is normally associated with the human mind (intuition, conceptualization, abstract imagery, etc.) is the exclusive purview of CB humans; these activities of the human mind traditionally have been associated with orders of priesthoods, schools of philosophy, universities, monasteries, prophetic traditions, think tanks, artist colonies, writer workshops, salons and every other node of creative thought that have appeared throughout the space-time continuum of human development.

Thus, it is typical to see groups of CB humans discussing *ideas* as their main contribution to the progress of their own colony, a la the Ionians in ancient Greece… and the work of producing the basic needs of the carbon-based, cyborg and the silicon-based members of the society of the colony. And, once again, as occurred about 500 to 100 BCE in the Hellenic Civilization, philosophy, the arts and letters are creating another vibrant human region in outer space.

Most important, however, is the continuing development of the human mind, which began as a seminal *hologram* during the Great Human Awakening Event; whose main function was to direct the colonization of Earth, but which now has matured to the point where it is now the driving and conceptualizing force in the colonization of the Solar System…

Chapter 35

Harnessing the Forces of the Cosmos

Ultimately, the colonization of any part of the Solar System by humans has depended on the ability to gain functional control of four major forces of the universe, in some fashion or another, and to the degree or extent required by the circumstances of the place and the moment...

Gravity, Electromagnetism, Strong and Weak Forces interact to establish the "rules of the field" which establish the bounds of all human activities in the Solar System...

From earlier narratives, it is evident that all the achievements of humans in pursuit of the mission to colonize the home planet solar system have been subject to the "laws" of the physical world and, therefore, that their efforts in advancing this agenda has been a manifestation of their *domestication* of gravity and electromagnetism and, implicitly, the strong and weak forces which operate within the then unappreciated atom.

However, the forces of gravity and electromagnetic were intuitively obvious to the earliest humans as soon as they descended from the trees and began to operate on the surface below them. Thus, they could no longer negotiate the attractive force of gravity in the same way they had done in the trees, where their longer extremeties could be used to swing from limb to limb (effectively utilizing a "slingshot" maneuver which would prove useful two million years later by astronauts in outer space.)

Indeed, on the surface of the planet, they would encounter the plenary downward pull of gravity and the "friction" impeded movement along the two-dimensional space. So, the Pleistocenes developed their skeletons and muscles – and renovated and enhanced the powers of their brains and sensory systems. Thus equipped, they began to stand upright, and to walk and run efficiently across the surface of the planet. In effect, they were managing (or at least negotiating with) the gravitational (or g forces) exerted on the surface of planet Earth.

These earliest humans also learned to utilize their muscles and their reoriented skeletal system to lift, push and otherwise finesse the lateral g forces on the ground, as they began to colonize the Tropic of Cancer Region. And, when their muscles were found to be inadequate for a particular task, they learned how to utilize the raw materials of the lithosphere and biosphere to fashion tools and develop techniques to multiply or leverage their innate capabilities. Thus, the overall experience of human development on Earth can be seen as a continuing series of effort to manage the g forces that are manifested in a given situation.

Humans also began harnessing the powers of non-human entities to enhance their inherent capabilities for managing gravity. This became more prevalent during the Holocene Period, when they began harnessing the strength of animals to do the *heavy lifting and pulling* of masses that could not be *manhandled*.

But even beyond utilizing the combined musculature of humans and animals as force multipliers, the Holocenes turned to the fashioning and employing inanimate objects to manage the g forces of planet Earth. Thus, they have utilization of such things as ropes and pulleys; poles and fulcrums; and a myriad of other such muscle enhancers to negotiate with the forces of gravity on the planet.

Now, in the age of space colonization, earthlings have learned how to utilize the macro force multipliers that are inherent in nature. Thus, for example, the natural force of gravity and angular tension are being used like cosmic magnets to add momentum to the velocities of low-mass artificial bodies. So, transiting spacecraft careen from one massive natural body to another, like steel balls in a cosmic "pinball" surface. Even the great explosive energy that is released by solar and cosmic eruptions is now being captured by powerful capacitors that are deployed at strategic locations at various orbits.

Surface Navigation

The Pleistocenes developed the first inborn systems for maintaining proper balance on the surface of a planet, as they descended to the surface of Earth. Thus, the eyes and ears, as well as the skin, adapted to perceive the external environment from the new perspective of ground-level, rather than from the vantage of the trees. They then developed the innate systems within the brain to process the data, and to convert it into actualizing information to muscles and bones via a reoriented neurological subsystem.

> *[It was only natural that the humans – who had internalized their separateness from and superiority over the natural world – would develop technologies to enhance the innate sensor-brain-neurological-skeletal systems to maintain the internal balance of the organism with respect to its surrounding environment on the surface of the planet.]*

However, the paradigm for managing surface gravity on Earth had to change significantly, as modern humans began their ascent, from the surface into the ascending strata of the atmosphere and, into the netherspace of the frontier between the stratosphere and outer space, and ultimately beyond Earth;s gravitational bond.

The "Escape Event" was historically preceded by the first attempts to manage the gravitational attraction of a planet in the 19th century CE, when Holocenes created balloons filled with lighter-than-air gas to transport human and non-human payloads into the atmosphere... followed by a continuing series of efforts to negotiate with planetary gravity in the third dimension; most recently through the use of aerodynamic and propulsion technologies.

Along the frontier between the planetary atmosphere and outer space, gravity could also be managed by the use of *jujitsu-like* maneuvers, which employ the angular momentum of celestial bodies, combined with bursts of artificial propulsion, to launch spacecraft into orbit, to maneuver spacecraft along desired trajectories and orbital pathways of interplanetary space, as well as within the gravitational wells of planetary bodies.

So, what humans have learned from their experience on the home planet, is that "surface navigation" does not necessarily involve movement along the surface of a planet; that is, it also can occur within the envelope of gases – the atmosphere – to the extent that it exists. In a sense, movement along a curved trajectory that mirrors the curvature of the celelstial sphere is also surface navigation.

Managing Electromagnetism

Human management of the forces of electromagnetism has occurred with two sets of objectives. On the one hand, humans have sought ways to shield themselves from the deleterious and dangerous effect of radiation. At the same time, they have also sought to harness the properties of electromagnetism to illuminate the darkness, to propel their vessels and as a means of teleporting information.

Even the Pleistocenes instinctively perceived that there are significant variations in the location and duration of direct sunlight on the surface of the planet which they were colonizing. They also were intuitively aware of the fact that certain objects – such as those of rocks – were hotter to the touch than others, such as wood and vegetation when the Sun shone on them... and that these various objects cooled – some more quickly and more thoroughly than other objects – when the Sun didn't shine on them.

They also perceived that the rays of the Sun had a sapping effect on them as they did work – especially during the time when it was directly overhead. And, they also noted that it was always cooler under the shade of trees or of hills. Therefore, it was natural that they began to developed ways to acquire shading from the effects of solar insolation, by using terrain and vegetation to shield themselves from the heat of the day, and by arranging their daily activities to take advantage of the diurnal pattern of sunlight.

Flash forward to the present time. Today, human management of the visible, infrared and ultraviolet rays of the Sun consists of finding ways and means to shield against them. They are simultaneously extending the hours of Sunlight, through the manipulation of flowing electrons. And, they are moderating the atmospheric heat that is generated by the Sun's rays, by "conditioning the air", by removing heated air and replacing it with artificially cooled air.

Now that humans are living in outer space, beyond the filter of any planetary atmosphere and the natural shield of a magnetosphere, they find themselves exposed to other phases of light and energy, along an electromagnetic spectrum of wavelength/frequency – which present a whole new set of challenges and opportunities for humans and their societies, mainly because there is no natural shielding mechanism out there. In other words, unprotected humans in outer space are exposed to the cell damaging effects of electromagnetic and nuclear radiation.

And, in outer space, sunlight and starlight is being perceived in new ways, as discrete packets of energy called photons, which have varying degrees of penetration and lethality and of variable size and mass that can have negative effects on humans in space. They are being generated by many sources, including our Sun, the galactic stars, as well as black-hole accretion disks, gamma-ray bursts and other like cosmic events which produce acute effects. Then there those subatomic energy packets that manifest as high-energy massive particles such as electrons, muons, protons and anti-protons.

There is also the perception among space colonists that the charged particle is the essence of *weather* in outer space which… while different in *kind* from the elements of atmospheric weather humans are familiar with on Earth, (and are discovering on some other celestial bodies), is *similar* in its *weathering* effects on materials. Therefore, radiation: in the phases of visible-light and ultraviolet; of microwave, infrared, and X-rays, and gamma rays… are elements of weather in outer space, which must be managed through a variety of technological negotiations, especially in places where there is no natural shielding from magnetospheres and atmospheres.

One or both kinds of "weathers" occur on the celestial bodies of the Solar System, depending on their shielding capabilities. To begin with, the weather that is related to the electromagnetic radiation and the radiation released by the decay of the nucleus of atoms encounter all celestial bodies equally, but not all radiation causes the same effect on the bodies: the bodies that have an effective magnetic envelope are shielded by that mass of charged particles; the stronger the field, the more efficacious the shielding effect. Those bodies that do not have magnetic protection are at the plenary mercy of the harmful radiations. A secondary level of planetary shielding from the effects of cosmic and nuclear radiation, as they "precipitate" on the surface of a body, is the field of gas, called an atmosphere – on bodies that have been able to formulate such fields.

Thus, on planets and the more vigorous celestial bodies that contain dynamos at their core, there are magnetic envelopes or magnetospheres and atmospheres that can subsume, dilute, deflect or otherwise mitigate the most harmful effects of cosmic and solar radiation. The negotiation with these forms of radiation involves the use of human technologies to enhance or supplement the natural shielding. And at a human level, shielding is achieved by using natural and artificial materials to block out the harmful radiations.

In those circumstances – in outer space – where there is no global protection against harmful cosmic radiation, humans have developed technologies to provide space colonists and astronauts with more community personal protective shielding from the radiation, by using protective spacesuits and protective covering on spacecraft and other structures.

Shielding also is an important consideration in the design and construction of space colonies, whether in orbit or on the surface of a celestial body. Natural shielding derived from rocks, soil and even water is used to protect the structures, their components and systems, the colonists, and other organic life from the harmful effects of solar radiation… especially in situations where there is no magnetic field or substantial atmosphere to attenuate the full force of solar energy incidence. Where the use of natural materials is not feasible, as in the case of the orbiting space colonies, artificial materials are being used to block, filter, or disperse harmful cosmic and solar radiation. *Smart* materials – which can change their molecular composition in real-time, to respond to changing radiation – are now being used to protect the exterior skin of space colonies.

Conversely, as space colonists have discovered more secrets of the electromagnetic spectrum; they have become intimately *familiar* with it – they have also developed a varied and powerful set of tools and skills for *domesticating* it… such as the use of the technologies of light and optics to observe phenomena, both distant and near, throughout the diurnal cycle… or the use of focused and enhanced light to observe phenomena; to transmit information; to transport energy; and to control the machines.

Managing Alien Environments

Once humans began to extend their colonization program into the interplanetary medium, they entered a truly alien natural environment; a place of desolation for earthlings, unlike the atmosphere on Earth. There is no oxygen and the gravitational ambience drains the blood from the brain and distorts its flow patterns to the heart and other vital organs. And, without artificial apparatuses to maintain the integrity of the human organism, the normal functioning of its internal systems, and to supply the brain and body with adequate doses of oxygen, the human organism will quickly die.

A similar environmental paradigm exists on the celestial bodies. On the planets and moons, there is insufficient oxygen. The atmospheric pressures differ in degree and so do the temperatures. In every instance, the temperature extremes exceed those of Earth's deserts and the polar regions. Without a special artificial life-support system, a human will die in a matter of seconds – from asphyxia or from exposure to extreme cold or heat, or from being "crushed" by the weight of the atmosphere on the surface of a planet.

Fortunately, there are analogs of environmental extremity on Earth, where scientists and engineers can develop the technologies that are needed to survive in outer space and on the celestial bodies of the Solar System. Thus, the natural environment of the higher altitudes Earth's atmosphere also requires the application of technologies to maintain the optimum level of oxygenation and the required balance between the internal pressure of the organism and the external ambient pressure. An appropriate balance is necessary to maintain both the anatomical and physiological integrity of the human body, to inhale and exhale naturally, and to maintain the circulatory systems in normal operation. It also is necessary for the brain to operate at optimal level.

Modern humans had already become intimately familiar with the dangers resulting from too much pressure on the organism when they began to venture into the lower depths of deeper bodies of water on Earth. There they became familiar with the phenomenon known as *the bends,* a potentially lethal biological event, which is caused by moving too quickly between extremes of atmospheric pressure. The opposite was found as humans entered the stratosphere, where the atoms of oxygen were too widely spread to be useful for breathing, and where there was insufficient atmospheric pressure to maintain, not only normal breathing, but also the integrity of cells, lungs, the heart and the brain, and the circulatory systems that transport oxygen and other nutrients to the vital organs.

Walter Gomez

SILICON-BASED HUMANS HAVE BEEN THE FIRST EARTHLINGS TO COLONIZE THE OUTER SOLAR SYSTEM HINTERLANDS...

Consider the vastness of the heliosphere, where even at the speed of light, (about 186,000 miles per second) a photon of light can take as much as 20 minutes to travel from Earth the edges of the Solar System. As a practical matter, the delay inherent in telecommunication of digitized information in space is problematic, especially in situations where anything approaching a real-time response time is required.

The time-delay phenomenon is even more challenging when payloads which have mass, rather than photons of light are being transported. One solution to this challenge has been to increase the travel-time of a transporter-payload (t-p) unit, to achieve some very small approximation of the speed of light... by minimizing the mass of the transporter and/or its payload, while maximizing the marginal propulsion. Unfortunately, in the real world of outer space, bodies with mass of any magnitude can never achieve even a microfraction of the speed of light.

Another approach to solving the problem of transporting massive t-p units over the vast distances is to deploy a series of relay transmission-enhancers to achieve adequate speeds. This gambit works well for transporting certain high-value goods, but it does little to serve as a routine method of transportation for payloads of any appreciable bulk. As a practical matter, however, the only conditions under which such technologies can be cost-effective, are those in which carbon-based humans cannot be transported to the remotest colonies – at the interstellar frontier – without expending massive inputs of technology, much less carrying out the prolonged and sustained work of constructing, activating, operating and maintaining such a vast network of relay stations.

So, the great distances between nodes of the Geospace Region also makes communication of raw data and finished information, too, a daunting proposition. Even at speeds approaching the speed of light, transporting packets of bytes, pixels and encoded photons via light beams through the heliosphere takes appreciable and significant time (from 2 seconds to as much as a 30 seconds).

The approximation of *real-time* communication between any two nodes, even with the most advanced technologies, becomes more problematic, if not impossible – as the distances between transmitters and receivers increase. And the longer the time-delay between transmission and reception of signals is, the more tenuous the sensor-processor-actuator loop will be. Therefore, the greater the length of the loop; the greater the need for operational autonomy at the various points within the loop.

On the other hand, for SB humans to perform these tasks, in those isolated, frontier colonies… they must be able to function with complete autonomy… with no direct control by any CB human entity… and they must also be equipped to survive and thrive in the physical environment of the outer heliosphere – where solar radiation is weakest, but galactic radiation is strong… where CB humans cannot function, without the application of massive shielding technologies.

Chapter 36

The Geospace Network

The traditional way of constructing human spatial networks on Earth, has been to first create the nodes (settlements, towns, cities, etc.); then to develop the linkages between nodes; and to thereby create an interconnected and interdependent spatial system. This narrative focuses on the nodes and linkages of the Geospace Region.

As a matter of context, consider that the earliest linkages on Earth were made visible by the paths created by human and animal legs, along which were transported *payloads* of ideas, information, and goods from one node to another. Later, during the Holocene Period, boats came into use as transporters across bodies of water, and these linkages were manifested by harbors and shipping lanes.

In the modern era, trains, planes, and motor vehicles are being used as transporters; and the rails, airports and highways are the physical evidence of these linkages. Additionally, the telegraph and the telephone, the radio and the world-wide-web have emerged to create another dimension of linkages and supporting infrastructures.

Thus, throughout Holocene history, the nature of linkages has changed in concert with advances in transportation technology. Also, some nodes have become more important than others, because of a comparative natural or cultural advantages. The comparative advantage can develop due to the presence of a natural resource such as water or mineral; a cultural magnetism (political or religious); or simply a serendipitous location that appears to be more attractive. Often the comparative advantage of a place simply appears in the mind of the humans as a *special place* (a war memorial, for example). In any case, it is a fact that certain nodes (places) become more important than others, as per various criteria, and develop a paramount relationship over other relevant places in a spatial system.

Viewed from another perspective, the comparative advantage of a node can be evidenced by the strength of its linkages to other nodes in the spatial system, including how many units of *payloads* it carries. Other indications of the strength of a linkage might be the magnitude and complexity of the physical infrastructure that frames it, such as the number airports, harbors, or train depots that dispatch and receive the automobiles, buses, trains, planes and boats that move between them.

Geospace Spatial Metrics

The concept of *distance* is an important aspect of linkages between nodes. It is more than an expression of physical distance along a line that connects two nodes; it is also a measure of *difficulty* – which can be measured and expressed in many different currencies. Within the Geospace Region, *real distance* is measured by the difficulty or cost that is involved in overcoming gravity and time; in transiting or transporting payloads (of matter or energy) between space colonies. Within the interplanetary medium itself, the main difficulty that defines *real distance,* is distance itself; but on celestial bodies, *real distance* is embodied in a variety of obstacles that require additional inputs of force, in search of the ideal, frictionless movement.

With respect to communication *distance* within the Geospace Region; it is defined by the time it takes for an electromagnetic transmission to reach its intended receptor; and in the case of interactive communications, it refers to the time it takes for an initial transmission to reach the intended receptor, and the time it takes for a retransmission to close the loop.

In practice, most communications in space are interactive in one form or another, and the great distances between any two places in the system, has required solutions to the *time-delay* problem that was discussed above. Another way of describing the communications paradigm is in terms of the input-processing-response (IPR) loop. On a micro-level, this refers to computer processor loops.

Another technologic response to *distance* in the heliosphere has been the development of a network of communication networks, which usually encompass some application of electromagnetic telephony or telegraphy. Currently, radio and laser-based technologies are the primary ones that are used to communicate at the approximate the speed of light through the interplanetary medium.

A major component of the overall cost of developing light-based communications within the Geospace Region is the deployment of hundreds of multi-ton mirror satellites in the various orbital zones – extending from the Cislunar domain, to the Martian and to Jovian domains… and even to the heliopause. But, regardless of the number of mirrors and other light-enhancing technologies that are deployed throughout the Solar System, there is still a maximum barrier to the energy of a light beam and, therefore, its maximum range. This barrier also limits the speed with which an object with any mass can travel through the region. So, even if it is possible to digitize matter, it still subject to the limitations of the speed of light.

The Light-Year

As is the case with all regions, the natural Solar System and the "artificial" Geospace Region can be appreciated at either a macro or micro scale. Each is manifested or measured by a "ruler" that is appropriate for a given mode of travel or transportation. Thus, the light-year, as the term implies, is the distance photons of energy can travel, through the quasi vacuum of space, and at nearly the speed of light. At the present level of human understanding of the laws of physics, the distance a beam of light (pure energy) can travel through space is approximately 63 trillion miles. In practical terms, that means that a beam of light can travel from one end of the Solar System to the other takes about 20 seconds to complete its voyage.

Conversely, because matter has mass, it cannot travel at any approximation of the speed of light. As a practical matter, this means that asteroids and comets, and spacecraft can only move at speeds that are less than the speed of light – depending on their mass. In the case of the latter, actual speed is also determined by the external force of artificial propulsion. Therefore, the speed of spacecraft is the product of magnitude of internal mass and the force of its external propulsion system. This means that space colonists and their machines cannot travel at anywhere near the speed of light, with the technologies of the present.

But they can utilize light beams (at any frequency within the electromagnetic spectrum) to project *avatars* of themselves or *bytes* of information to other places within (or beyond) the Geospace Region. Light beams also can be used as measuring devices to determine the distance between a given transmission point and receptor. As an example, knowing that light beams travel through space at 186,000 miles per second provides a measuring device for calculating distances between any two celestial objects, by measuring the *time* it takes for a beam of light to travel between the two points in space.

Meanwhile, continuing advances in the science and technology related to the deconstruction and digitization of organic and inorganic matter, will soon permit the transmission of matter as encoded and encrypted information, via photons at nearly the speed of light through space… and the reconstruction of the matter at a targeted reception and reconstruction point.

Thus, it appears that humans are approaching a time when they will be able to transmit, not only bits of data and pixels of imagery, but also atomic and subatomic particles of matter, throughout the Geospace Region, at nearly the speed of light. This alone will reduce the size of the *virtual map* to a navigational reality, in which the maximum distance between any two points within the human space region will be measured in minutes and hours.

Familiarity with the reality of outer space also has been extended to the development of routinized approach in conducting the space colonization. More than a century of space exploration has produced an extensive archive of data and information, and a huge constellation of algorithms – all of which have been added to the Common Archive. This cloud of experience and knowledge, of know- how and skills has been a powerful tool for conducting the

So, it is not surprising how quickly the space colonization program has been implemented, and how rapidly the space colonies have been deployed in every zone of the Solar System. Consider that the first constellation of space colonies was constructed throughout the low-Earth orbital zone in the year 2020 CE; that the Lagrangian space colony constellations were deployed and in full operation at the end of that decade; and that by 2060 CE, the outer space colonies of the Kuiper Belt had been integrated into the Geospace Region.

A major reason for the acceleration in the pace of colonization is the enhanced power of the human mind, which has enabled significant advances in the way space humans can access knowledge, technology and skills related to any subject by using powerful *search engines* that are integrated within the mind, to access the Cosmic Database – directly and interactively – for *information and guidance* on any subject.

So, the Cosmic Database now contains a comprehensive compendium of the of human experiences that have accumulated over the past two million years. These have been digitized… inputted, scanned and streamed onto the millions of hard-drives and other computer storage devices that are attached to the internet systems of the Geospace Region. The heart of this powerful information system are the *algorithms* (interactive instructions) that guide the search engine in its search for all possible information related to the search request.

Chapter 37

The Geomars Transportation System

The low-mass strategy for colonizing space is manifested in the blueprint for the development of the Geospace Transportation System (GTS): it outlines the construction of a comprehensive transportation network of orbital "low-mass highways" in the gravitational field of the heliosphere. The terminals of the GTS include the primary cosmodromes that serve Earth, the Moon and the Lagrange Zones now; the cosmodromes that are being planned for Mars and its moons; the staging platforms for future colonies; and a subsidiary network of secondary spaceports, which serve the space colonies, the mining centers, and the specialized bases within the Geospace Region.

The primary function of these cosmodromes and spaceports is to serve as transshipment points, but they also provide safe harbor during cosmic and solar eruptions. The cosmodromes also contain maintenance facilities where spacecraft can receive operational maintenance or major refitting. Parts and equipment are stored in warehouses, but nanotechnology and 3D printing technology is also available to create virtually any part or system that is required to create or renovate a spaceship of any size or complexity.

The cosmodromes also provide space tug services for more massive spacecraft, to facilitate their launch or landing activities. They also offer specialized support services to cargo spacecraft, such as those that are carrying water-ice or ores and other bulk materials from asteroid and comets to processing centers. And, the cosmodromes on the Lagrange Zones provide refueling, resupplying and refitting support to passenger spacecraft that are embarked on very long voyages to the Martian and Jovian Subregions.

Aside from its general mission of providing linkages between space colonies, cosmodromes now provide an assortment of specialized services to the newly-developing mining sites on the asteroids. Indeed, the initial main function of the GTS was to haul semi-processed mineral ores from the asteroid mines to Earth for final processing, and for use as raw materials by manufacturing centers there. But now, these raw materials are initially processed at the asteroid extraction sites and are then shipped as semi-finished materials to the new manufacturing centers on the space colonies, as well as on Earth.

Meanwhile, the commercial and passenger aspects of transportation continue to be an important component of the GTS, as people travel routinely for business and pleasure, and as merchandise is bought and sold by people on the various space colonies and on Earth.

An important component of this new transportation system is the space plane, whose design is based on the NASA Space Shuttle. This prototype "spaceplane" transported astronauts and cargo between Earth and the International Space Station during the early years of exploration and scientific study of the Solar System.

An auxiliary to the spaceplane is the space tug; it is a heavy-duty propulsion platform that is designed to facilitate spaceplane launchings and landings through the deeper strata of gravitational wells on the more massive celestial bodies. Space tugs also are used to provide an additional boost to launching spaceplanes from the angular velocity of a spinning celestial body.

But it was only with the construction of the orbiting spaceports, and the construction of landing strips over the entire surfaces of celestial bodies, that a full range of transportation services is available throughout the Geospace Region. Now, even at the most remote bases on asteroids and comets, spaceplanes can receive and launch – in a way that is reminiscent of bush aviation or naval carrier operations on Earth.

By way of summary: The spaceplane has become the key component in the development of a comprehensive Geospace Transportation System. Its versatility as a platform now offers the opportunity for all space colonies and specialized space bases to interact with any other node in Geospace. The spaceplanes are inexpensive passenger and cargo vehicles that can fly multiple sorties; withstand the stresses of launches and recoveries within the deepest gravitational wells; and operate routinely in the most extreme environments of the planets and other celestial bodies and of the interplanetary medium.

The spaceplane also displays its toughness in its designed ability to land and takeoff from Earth, Mars, the Moon and other celestial bodies; on any kind of terrain, thus giving it the capability to operate on the most ragged asteroids and comets as well. It also can perform every kind of rendezvous maneuver to load and offload payloads from any orbiting colony or even the smallest orbiting outpost.

The key to its ruggedness is in the materials that are being used to fabricate its structure and operating systems, and in the redundancy that is built into them. The materials are specially designed and fashioned in the microgravity environment on space colonies, to make the spaceplane rugged enough to operate efficaciously even in the most extreme physical environments of the Solar System.

Equally important is the increased efficiency of its ion propulsion systems, which provide more power, with less mass. And, the spaceplanes are provided with a host of performance features, which include thrusters and flaps that are part of an integrated control system, and embedded sensors and actuators that can detect and automatically respond to issues that affect the performance of the spaceplane.

Specialized versions of spaceplanes are designed to perform a variety of missions. In some situations, heavy rocket launch systems are combined with spaceplanes, in a configuration that was pioneered by NASA with its space shuttles. But these modern hybrid systems are more advanced than their space shuttle precursors, in that they can take off and land on airports, and are also able to alight from and hover over space colonies and celestial bodies, without the need for heavy missile launch systems.

Other spaceplanes function like tugboats, which are specially configured to pull or push space barges which transport commodity bulk products such as ores from the asteroid mining operations to other points in the system for further processing or the manufacturing of industrial and consumer products.

Still other spaceplanes are designed for the mission of maintaining the security of the space lanes, in the same fashion as the naval warships that patrol the shipping lanes on the oceans of Earth. Thus, the spaceplanes patrol the space lanes to protect against a variety of threats, including those that are still posed by rogue nation-states or non-governmental pirates or terrorists. These threats include Earth-based missile attacks, orbiting mines that can damage or even destroy a satellite, hostile spacecraft within a given orbit, and even attacks from gunship-satellites. It also should be noted that the fact that there is a need for space planes to provide security to the region is an indication of how the assets in space have grown in number; the extent to which they have become essential to the survival of the human species in space; and the self-evident exposure of these assets as targets from rogue nations or terrorist groups on Earth.

Chapter 38

Nanotechnology in Space

Pursuant to the overall strategy of minimization of mass in the colonization of the Solar System, it is essential to minimize the size, weight and mass of every structure; the materials used to fabricate each component of the structure; and the internal systems of every structure.

Minimization of mass also is an essential basis for the development of the transportation system in Geospace. It is inherent in every cosmodrome and spaceport; in the spaceplanes and spacetugs; and in every other component of the network. It is evident in the everyday work of lifting and manhandling objects; and of turning knobs and opening hatches. Minimization of mass is important in carrying out every phase of space operations.

Or, alternatively, one can increase the force that is applied to move a payload through interplanetary space, or to maneuver objects in within space colonies. Most commonly, the maximization of force technique is employed to boost spacecraft out of a gravitational well, or to provide a graceful re-entry into it. "Big-Boost" techniques are also used to provide extraordinary power to spacecraft: to move to another point on the same orbit; to transfer to another orbit; and to rendezvous with another object in space.

These techniques of modulating propulsion forces to move massive objects were first used during the era of "big missions" of exploration, and were usually employed in conjunction with "slingshot" maneuvers – which used the angular forces of spinning planets to augment the internal propulsion force of the spacecraft. Now, in space colonization operations, the challenge of transporting payloads in space is now more likely to be dealt with by minimizing the mass of the spacecraft-payload unit and maximizing the internal propulsion force.

A crucial aspect of the mass-minimization strategy is micro-miniaturization. This technological approach utilizes molecules as the "building blocks" for creating materials and components with the least possible heft and mass. The reasoning behind this approach is that the combination of lower-mass payloads, spacecraft, and propulsion systems will require lower magnitudes of propulsion force and, therefore fuels.

To this end, the overriding imperative in the movement of inert mass in space is to minimize the size and weight of the objects that are being launched and propelled through space. In this context, "smaller" means reducing, not only the size of the spacecraft-payload unit, but also the total mass of the matter it contains. In other words: micro-minization. This has been the great leap forward in conveyance technology that has enabled humans to operate efficaciously in the realm of distances between nodes that are measured in terms of Astronomical Units (about 90 million miles) and in light-years (about 6 trillion miles).

Nano-Technology

Nanotechnology – refers to the design and fabrication of materials and components, as well as whole systems, at the scale of a billionth of a meter – is the essential ingredient in space transportation. It is also the essential mechanism in the overall process of minimizing mass in space colonization.

Nanotechnology also has made possible the development of comprehensive medical facilities onboard space colonies. Specially-programmed SB medics now can provide immediate diagnoses of health issues and apply healing at the cellular or genetic level, thus obviating the need for traditional surgery which would incapacitate a colonist for some time. In a very real sense, all medical treatment has been reduced to the level of the common cold.

As indicated earlier, nanotechnology is used aboard space colonies to manufacture nanostructures. These engineered nanomaterials can be constructed either by a top-down approach, where large material is reduced in size to nanoscale particles; or through a bottom-up approach in which larger structures are constructed, atom by atom, or molecule by molecule.

Thus, space colonists have become *familiar* with the powers of nanotechnology and its capabilities in the fabrication of specialized materials that are light enough to minimize mass and maximize tinsel strength, to withstand the stresses imposed by the natural environments of the interplanetary medium, and of the various celestial bodies of the Solar System. Additionally, the alchemy of nano-technology and nano-based materials that are produced in a micro-gravity environment, has produced metals and plastics that are lighter and stronger than any found on Earth. In summary, space materials greatly reduce the mass of a spacecraft to reduce launch costs, and to lessen the cost of deploying and assembling the infrastructure of the space colonies. They are the building-blocks for constructing lighter and stronger structures that still meet the demanding specifications of the outer space environment.

And, it is the case that nanotechnology and nanosensors are a vital to the strategy of minimizing the mass *footprint* of space colonies on the gravitational field. These applications have heightened the probability of successful human settlement throughout the Solar System, because it increases the variety of technological responses that can be applied to a greater diversity of environmental challenges. It is a major reason why space colonies are now operating satisfactorily, even in the most extreme environments of the interplanetary medium and planets and celestial bodies.

3-Dimensional Printing – is the corollary technology that is available for fabricating materials at the level of nanometer; with this capability, space colonies can design and construct virtually any material, component and system that is needed on the space colony, thus mitigating the aggregate quantity of mass that must be transported – from Earth, the Moon or Mars.

And, as a general matter, the cost of transporting materials and goods over such cosmic distances, makes it imperative to create systems and equipment in space with the least overall quantity of mass. The colonies and other facilities themselves must be designed to leave the softest *footprint* on the gravitational mesh of the Solar System.

Nano-Sensors

Another important element of the nano approach is the development of artificial nano-sensors, which enable the human eye and other senses to observe, study, measure and manipulate nanomatter and nanoprocesses. One such artificial enhancer is the electron microscope, which provides the visual acuity and resolution to pinpoint the behavior of single atoms and their subatomic components. It also enables scientists and engineers to measure magnetic properties of materials with atomic precision. And, with this power of precision, scientists and engineers can manipulate the magnetic properties of nano-structures for applications in space colonization.

Another example of such nano-sensing capabilities is the scanning electron microscope, which uses electrons instead of light to form an image. It is one of the most powerful contributions to the colonization of the solar system, especially when it is used as *in situ* sensor that is integrated with planetary landers and rovers. In such applications, this tool has been invaluable in providing the ability to do real-time, *in situ* analysis of minerals, soils and other samples, which are drawn from the atmosphere and surface of planets, moons, and other celestial bodies. It has been particularly useful in developing asteroids and comets as readily-accessible sources of natural materials for orbiting space colonies.

At one end of the scale, this advanced microscopic system enables the observation of the Solar System at the subatomic level; at the micro scale, provides the power to view the activity of electrons within the atom with the highest resolution, discrimination, and visual acuity. The nanoscope is the technological key to the world of the atom which previously had only been intuited by the powers of the human mind, as inspired by the Cosmic Mind.

One special feature of the nanoscope is that it is an *active* sensor. That means that, unlike the *passive* versions, it emits its own "illumination" of the object it is observing; it emits a beam of electrons on the tiniest objects, and uses the reflected light from these electrons – rather than visible light – to generate a much higher-resolution, more detailed image than ordinary visible-light sensors.

The nanoscope is another example of humans learning to use visible light and other phases of the electromagnetic spectrum, not just passively, but actively to study phenomena. It is another example of increasing *familiarity* with the forces of nature, not unlike the domestication of fire by the way Pleistocenes some 1.5 million years earlier.

Nanosensors also are being used in the pursuit of low-mass strategies in the design and manufacturing of spacecraft; they are embedded into both the outer and inner surfaces of space colony structure; where they can detect, measure and process stimuli they receive from the outer environment and the inner environment within the space colony – and then take appropriate responses to maintain the normal operation status of the space colony.

Chapter 39

Quo Vadis?

The future of space colonization lies in the hands of the silicon-based humans…

Consider that the only way to colonize the outer regions of the Solar System, where immense distances make effective control by controlling entities on Earth or Mars, or even Neptune virtually impossible… is to imbue SB humans with the same powers of the brain and mind that previously were only possessed by CB humans. This represents the final step in imbuing the SBs with the same degree of *humanity* as the Holocene humans.

This means that both types of humans are now equal partners in any dialog that deals with planning, designing, activating and operating space colonies. Now CB and SB humans can work together on the macro activities of conceptualizing and planning; because both can do the work that is usually associated with the right-side brain – such as the detailed auditing of processes and the rigorous re-evaluation of algorithms, which is aimed at minimizing the miniscule "bugs" that can derail a project, or an entire mission.

Silicon-based humans are now also imbued with the same degree of *human sentience* and environmental awareness as their carbon-based colleagues and, like their CB predecessors, the SBs now possess the gift of *free will* and situational discretion… which enables them to respond, in real-time, to any unexpected development in situations where the empirical feedback loop is too long for effective control by an external entity.

Similarly, both types of humans have equal faculties for symbolic reasoning – they possess the same level of acumen to formulate ideas and concepts, and to deal with the abstract constants and variables... those that are expressed by strings of alphanumeric symbols and mathematical formulae... and which are variously combined to form words, phrases or sentences and paragraphs, in any syntax or specialized language, at within the context of any situation.

The vast storehouse of data that are stored in all the databases of the Cosmic Cloud are available to both CB and SB space colonists, as they continue the short-count and long-count processes of space exploration and colonization. And, all humans are privy to the universal reservoir of algorithms, logic and calculus that have been used to solve even the most intransigent challenges of space colonization that have been used effectively by their predecessors. At the same time, they realize that, for any general rule, there can be exceptions; very little is simply true or false, and that all algorithms must be constantly tested for their efficacy in changing situations and environments.

Consider that, because of the distance-factor and the consequent limits on effective two-way communication, SBs are now solely responsible for carrying out the colonization of the outer frontier of the solar system: the frontier zone between the interplanetary medium and the interstellar medium of the Milky Way galaxy. There the SB colonists must operate independently of any controlling entity that is based even on the nearest outpost (L1 Lagrange Zone of the Sun) – to develop long-term objectives and to set subsidiary goals; and to develop the ways and means to achieve them.

In practice, this means that SB humans on the heliospheric frontier now possess the plenary power to assess their needs and to deploy the resources that are needed to a desired short-count or long-count objective. Thus, they have full discretion in developing the human resources on each of their colonies; to deploy teams consisting of both SB humans, and an assortment of robotic workers and bots to enhance the efficacy of their efforts; and they can develop algorithms for harnessing the energy, and the knowledge and skills of all colonists within their domain to achieve immediate goals and ultimate objectives.

The upshot has been that the SB colonists have been quite successful in developing the algorithms to solve an array of problems related to the colonization of their frontier domain; and they also have developed a greater facility in the use of emerging algorithms and the application of swarm intelligence to determine the best courses of action when presented with challenges for which there is no efficacious "off-the-shelf" algorithm or paradigm.

SBs are also capable of periodically ascertaining whether the real-world matches established predictive models, and whether to update them. Beyond that, SB humans can resolve problems they encounter even where there is no established model or algorithm available; and they can create new predictive models to either replace obsolete ones or to supplement them. And, in acute and totally *alien* situations, they are equipped to quickly identify, classify and diagnose the problem; and to apply the appropriate corrective measures.

This ability to adapt and to make changes to the internal systems and structures – in response to changes in their external environments – is as crucial for silicon-based colonists on the interstellar frontier, as it is to carbon-based colonists in the Cislunar, Martian, and Jovian colonies: even more so, because they are practically on their own, without any backup from the colonies of the inner Solar System.

Out in the hinterlands of the heliosphere, the gravitational hegemony of the Sun diminishes, and the environment becomes increasingly galactic; the solar wind loses it force as it is confronted by the interstellar fronts and waves ripples caused by the life-cycle events of stars. It is the place where dwarf planets and swarms of comets create local perturbations… This is the *back to the future* zone, to which the cosmic artifacts from the *paleo-history of the construction of the Solar System* have been relegated.

The Planetary Project

Even as the campaign to colonize the entire Solar System with thousands of orbiting space constellations continues, there is an ongoing, long-term plan be developed to colonize as many of the planets, their moons, and even some asteroids. It is a project that began in the late 20th century CE, on Earth, and has continued throughout the period of deployment of orbiting space colonies in the 21st century. And, it will likely continue, on various tracks of purpose and capabilities for many centuries to come.

The Earth's Moon was the first candidate for space colonization for various reasons. To begin with, humans had been landing on "Luna" since 1969 CE; therefore, they had developed considerable experience in planning two-way missions (sorties) to the body and had routinized the optimum trajectories and rendezvous for landing on specific sites, and relaunching at the appropriate time to recover on Earth. Another comparative advantage that the Moon had over the terrestrial planets was its closeness to Earth, both in terms of transportation and communications. Its relatively low gravitational well was another factor in its favor as a candidate for establishing human space colonies on its surface. But, after due consideration and practice, it was decided that the Moon would be more difficult to terraform than either Mars or Venus, the next two candidates for permanent human colonization; it would be more useful as a logistical center for Cislunar operations and as a source of raw materials for Earth and the orbiting space colonies.

Venus has potential as a platform for experimenting with atmospheric space colonies on some of the Gas Giant planets that lie beyond the Main Asteroid Belt, but it too is destined more as a potential source of chemical elements, for processing and utilization on the orbiting space colonies, than as a place for human colonization.

Mars has been selected as the best candidate for long-term terraforming and as a site for permanent, large-scale populations of earthlings. But the project is long-term; centuries long, and it will involve capital investment that can only be provided from the cumulative profits from the operations of the orbiting space colonies.

Summary Conclusions

One of the major goals of colonization of the Solar System is to develop a *Homo spaciens* civilization everywhere…

That means, that all forms of humans in space: carbon-based, cyborg hybrid, and silicon-based humans will someday be able to not only survive, but thrive; to reside permanently as denizens of every part of the heliosphere and on every planet, moon, asteroid… in their naturally-evolved condition.

However, the reality of the present is that carbon-based humans have not yet achieved the status of Homo spaciens because the organic mutations and adaptations have not progressed to the point where CBs can live and work in space, or on a celestial body, without significant inputs of technologies and techniques. Nor have the cyborgs achieved the level of Homo spaciens – although they are further along in their evolution towards that status because of the relative efficiency of artificial "mutations" and the consequent advances in adaptability to some of the more "alien" environments of the Solar System.

Therefore, as matters stand now, and in the foreseeable future, only SB humans can carry out the colonization of this outer zone of the Solar System, because of the exorbitant cost and virtually insurmountable difficulties of any sort of technology "fix" to enable CBs and even cyborgs to live and work there.

The natural environments at the edge of the heliosphere are even more alien to carbon-based earthlings than those of the inner, terrestrial sector, or even the those that exist within the Jovian sector of the Solar System. Thus, they present a variety of exceptional and extreme *challenge-environments*… in which interstellar particles from novae are charged particles of combined nuclear and cosmic origins… which gives them more energy and, therefore, greater penetrating power… which makes them more lethal to organic cells and inorganic molecules, in the immediate and over the long-term.

Only the silicon-based colonists can propagate out beyond the terrestrial zone of the solar system, where the solar winds begin to interact with the interstellar fronts… where the gravitational and magnetic fields of the Sun and the inner planets became more tenuous, and where the magnitudes of space and time simply are too great for timely human communications. In these circumstances, the colonization program can only be executed by silicon-based humans.

Only the SB humans can live and work in extremophilic outer space without the need for *spacesuits* or any other protective capsule. The lack of oxygen and ambient pressure have no effect on their anatomical or physiological integrity; nor on their psychological stasis; and their materials are immune to the corroding or mutating effects of all forms of irradiation by charged particles. *In other words, the SB humans are the most efficacious beings for carrying out the colonization of the solar system – and beyond.*

The Sb Brain

The SB human brain is a virtual replication of the modern Holocene human brain; virtual in the sense that it is not an exact copy of the CB brain. Instead, its form can be likened to an ethereal, cloudlike distributed constellation of logic and memory nodes, which are interconnected and interrelated by dynamic electro-chemical processes.

The SB *brain* – at the time of its creation – downloads all the memories and associations of the Common Experience that have accumulated since the Great Human Awakening Event. At the same time, the SB entity receives algorithms that are necessary to interpret, validate and process the inputs from the Common Experience... which includes the accumulated knowledge and skills of the CB humans.

SBs are equipped with sensory systems that transcend those of the carbon-based humans. Their faculty of *sight*, for example, enables them to apprehend not just visible light, but also the other portions of the electromagnetic spectrum, such as in the infrared and ultraviolet phases. They also can detect and process harmful, destructive high-energy electromagnetic and nuclear decay radiations and respond to them with their enhanced immunity system. And, SBs have an innate ability to detect and process sounds, chemical *odors*, and even minute changes in their relevant gravitational and ambient pressure environments.

The SB brain is a powerful *perceiver of data, processor of patterns, and memorizer of events*. Each pattern recognizer within the brain contains about 100 million artificial nodes (neurons), equal to that of the CB brain, but because of the more efficient actualization of connectivity, SBs can recognize many more patterns than their CB counterparts.

And, by the year 2049 CE, silicon-based humans had developed ways to replicate... They have become the new *Homo spacien*, which will be the dominant species within the Solar System, and beyond...

> *[Just as the anatomy and physiology of Homo erectus adapted to changes in their local physical environments, so too is the architecture and operating system of the SB humans in space is updated, to respond to changes in their relevant physical environment, with a variety of programmed reparative responses.]*

Epilog

As I, Josef Herzog, write this narrative summary of human history in the year 2060 CE – 5 billion years after the birth of the Solar System; 2 million years after the Great Human Awakening Event on Earth; and in the 40th earth year of human space colonization – earthlings have developed the necessary knowledge, skills and tools for completing the Cosmic Imperative of colonizing the Solar System.

The most important element throughout this endeavor has been the human mind, which has been developing as a separate, virtual conduit for accessing the Cosmic Mind during the past 300,000 years. Its maturation and assumption as the directing agency in executing space missions of exploration and colonization occurred only 5,000 years before the present. Since then, it has continued to develop independently of the brain, in much the manner as the Artificial Intelligence machines had begun to do in the latter decades of the 21st century.

The instructions that have been transmitted by the Cosmic Mind has enabled generations of human species to overcome cosmic and planetary obstacles to their efforts to colonize planet Earth; in the case of the Pleistocenes this was manifested by the *times of trouble*. They began 300,000 years ago, when global warming caused the melting of the ice caps and glaciers, and the consequent global flooding caused a reconfiguration of the surface of the planet. By 30,000 years ago, the Pleistocene humans had been replaced by another, more advanced generation of Holocene humans who had a more sophisticated brain and maturing mind, greater powers of intuition and reasoning, an ability to communicate orally and symbolically, and their higher level of social intelligence. With the context of these gifts, the Holocenes would develop a new lifestyle of settled agriculture; one that was more effective as a survival strategy in the reality of the planetary environment. It would prove efficacious for the survival of the human species for thousands of millennia.

However, by the 19th century, the Holocenes had created another paradigm for colonizing the planet, one which involved an even greater separation of humans from nature. It was during the age of industrialization that humans began migrating from the traditional food-production nodes to the burgeoning nodes of manufacturing. It was a time when humans finally began to perceive the entire globe as a "planetary mine" from which to extract the raw materials needed to feed the growing industrial centers. And, increasingly, the industrial centers became, not only producers of consumer goods, but creators of more lethal, "better" tools of war.

The Holocene became so successful in their efforts to extract the raw materials from the Earth and, in the process, inflicting abuse on the natural environment, that they by the middle of the 20th century, they had reached the tipping point where there would no longer be enough resources to sustain the growing population. And, the densities of human populations – which were flouting the strategies of dispersal that had been followed in the millennia following the Pleistocene *time of troubles* – would not be able to survive a strike by an asteroid, a mega-volcano eruption, a thermonuclear war, or even a pandemic disease.

And indeed, the Holocene *time of troubles* began with a confluence of global warming; which not only rearranged the ratio between the waters and the lands, but also increased the amount of water in the atmosphere; which caused greater insect and bacterial activity; which caused global destruction of crops and enervating disease in humans; which created more stresses on populations; which generated an endless cycle of wars; which culminated in the use of thermonuclear Armageddon…

Geospace 2060

The New Diaspora

Ultimately, the Holocenes made the decision to begin evacuating the planet during the Holocene *time of troubles*; they took the first steps out of Earth by learning to operate in the stratosphere first, and then by escaping out of the gravity well of the home planet.

Airplanes began the process of dispersal into outer space, as humans ascended as high as the stratosphere, the very frontier with outer space. Spaceplanes conquered the transitional space between the Earth's atmosphere and the near-earth space that extended to about 200 miles above the earth's surface. This would be the place where humans would establish the first extraterrestrial colonies.

Spacecraft extended the frontiers of human colonization beyond the near-earth space as colonies were established in the Martian and Jovian domains. And, spacecraft then reached the limits of the Sun's solar winds as they flew past Pluto, beyond the Kuiper Belt and through the heliopause: the outer boundary of the heliosphere. It also is the outer frontier of Geospace Region.

A Stepwise Strategy

The strategy for colonization of the solar system has followed a stepwise strategy in the deployment of space colonies; and in conjunction with the strategy of deployment of space colonies in low-mass places. Thus, the first generation of human colonies in space would be deployed within the interplanetary medium, rather than on a planet or other celestial body.

So now the first phase of the mission to colonize the solar system has been completed... The first constellation of orbiting colonies has been deployed throughout the zone which lies within the area encompassed by the Earth and its Moon, and Mars – the Cislunar Subregion. There are now space colonies that are orbiting the Sun, Earth, and Mars. Many are located within Lagrange Zones, which have proven to be ideal places in which to establish the logistical and other support facilities, for existing constellations of human colonies, and as staging centers for the deployment of new colonies throughout the Martian, Jovian and Plutonian domains.

In support of these space colonies, there are thousands of artificial satellites, laboratories, observatories, security bases, way stations, cache depots, mirrors, transportation and communication relay stations, and a host of other specialized earthling nodes are in a variety of orbits around the Sun or other massive bodies.

The latest steps in the development of the Geospace Region has focused on the establishment of colonies on the "landmasses" of the Solar System: the planets and their moons. The orbiting colonies will continue to remain the most numerous and important nodes within a geocentric spatial system; Earth will continue to be the most important node, the most important point of origin and destination for all transportation and communications within this subregion of the overall Geospace Region.

But now the martian domain is also emerging as a Central Node in the developing subregion whose subsidiary nodes are the bodies of the Main Asteroid Belt and its moons. Jupiter is becoming another major node, whose great distance from the Sun and its own powerful gravitational attraction gives it hegemony over all the planets and the other celestial bodies whose orbital behaviors are primarily influenced by their distance to the combined gravitational influences of the Jovian Subregion.

So, now the human colonization of the Solar System is beginning its ultimate phase; one focusing on the development of "sedentary" colonies which will be deployed either on, or under, the surface of massive bodies, depending on the atmosphere, lithosphere and hydrosphere characteristics that is found on each body.

In a sense, the colonization of the planets will represent a return to the past on Earth when the Holocene planetary colonists began their quest to establish settlements on a little-understood planet.